PRAISE FOR SHIREEN DODSON'S

The Mother~Daughter Book Club

"This is a book that was waiting to be born, an angel book sent to give mothers and daughters hope. . . . *The Mother~Daughter Book Club* is about community, about adults supporting each other and helping other people's children. It's about communicating values to children in ways that work. It will help mothers and daughters connect in new ways and save many Ophelias from drowning."

—Mary Pipher, author,
Reviving Ophelia

"Provide[s] mothers and daughters with a wonderful opportunity to strengthen the bond between them."

—Tipper Gore

"*The Mother~Daughter Book Club* is an idea that is so wonderul, so full of common-sense goodness, why has it taken so long for it to be born? This could easily be the catalyst for a national movement."

—Jim Trelease, author,
The Read-Aloud Handbook

"Dodson's book is full of contagious enthusiasm and practical suggestions."

—*People*

100 Books for Girls to Grow On

- ✦ LIVELY DESCRIPTIONS OF THE MOST INSPIRING BOOKS FOR GIRLS
- ✦ TERRIFIC DISCUSSION QUESTIONS TO SPARK CONVERSATION
- ✦ GREAT IDEAS FOR BOOK-INSPIRED ACTIVITIES, CRAFTS, AND FIELD TRIPS

Shireen Dodson

HarperCollins*Publishers*

HarperCollins books may be purchased for educational, business, or sales promotional use. For information, please write: Special Markets Department, HarperCollins Publishers, Inc., 10 East 53rd Street, New York, NY 10022.

Designed by Ann Gold

FIRST EDITION

Library of Congress Cataloging-in-Publication Data

Dodson, Shireen.
100 books for girls to grow on / Shireen Dodson.—1st ed.
p. cm.
ISBN 0-06-095718-2
1. Young adult fiction, American—Bibliography. 2. Young adult fiction—Stories, plots, etc.—Examinations, questions, etc. 3. Children's stories—Stories, plots, etc.—Examinations, questions, etc. 4. Girls in literature—Examinations, questions, etc. 5. Young adult fiction, English—Bibliography. 6. Children's stories, American—Bibliography. 7. Children's stories, English—Bibliography. 8. Girls in literature—Bibliography. 9. Girls—Books and reading. 10. Best books. I. Title.
Z1037.D64 1998
[PS374.G55]
016.813'508092837—dc21 98–27606

---

98 99 00 01 02 ❖/RRD 10 9 8 7 6 5 4 3 2 1

To the influential matriarchs in my life—
Charlotta Colley, Felice Dodson,
and Lillian Spears Carter.

Acknowledgments

100 Books for Girls to Grow On could never have been written without the interest and support of many individuals. I'm especially grateful to the mothers and daughters of the original mother-daughter book club, who continue to inspire me. Jewell Stoddard at the Cheshire Cat bookstore in Washington, DC, has been an enthusiastic supporter of mother-daughter book clubs, a discriminating reader, and an invaluable guide when it comes to selecting books. I'm grateful to Louise Colligan, Nancy White, and Merri Rosenberg for their essential contributions; to Mary Freeman for her expert reading; and to the rest of the team at Seth Godin Productions, especially Lisa DiMona, Sarah Silbert, and Robin Dellabough, for bringing their considerable energy and heart to the process. At HarperCollins, my sincere thanks go to Megan Newman, Susan Weinberg, Jennifer Hart, Jane Beirn, and Jessica Jonap, for efforts beyond the publishing norm. My family and friends deserve accolades above all, for I truly couldn't have managed any of this without them.

Introduction

A few years ago, as my elder daughter edged her way toward adolescence, I felt a need to strengthen our connection before it had to weather those turbulent waters. In a moment of inspiration, I hit upon the idea of starting a mother-daughter book club. Nine-year-old Morgan and I invited several other mothers and daughters to join us once a month to talk about a book we had read. That simple idea brought Morgan and me closer together than I could have imagined.

Our club has been a great success. It fosters communication between mothers and their daughters. It forges friendships. And it's fun! We are enjoying year three, and the girls still look forward to meetings. They love the chance to talk about issues, especially the issues that would be tough to discuss with their moms outside this "safe" haven. The moms have an understanding that what's said in the book club stays in the book club. No one can be punished or chastised later on about something she revealed during one of our sessions. We mothers want to hear everything our daughters have to say.

The first year, our book discussions lasted 45 minutes. Not usually less, never more. Year two, they were an hour or so. We're now up to an hour and a half of strictly book discussion—the socializing and refreshment part is on top of that precious hour and a half. As we've all gotten closer, we're more apt to get into real-life discussions, and not confine our talk to the plot or characters in that month's book. The girls trust each other. Shy girls blossom. Quiet girls become outspoken. Everyone feels enough trust to participate.

Another marvelous outgrowth of our club is that whole *families* have become close. We do things as couples, or with the children. In fact, my family is renting a vacation house with one of the families from the book club for a week this coming summer.

Soon after our group started meeting, I knew I had learned something that was too valuable to keep to myself. I wrote my first book, *The Mother-Daughter Book Club* (1997), to let other moms in on my secret. That book describes in depth how we started our club and how readers can start clubs of their own.

I've been on the road talking about that first book ever since. Several major television appearances later, I continue to get calls for interviews from all over the country. Almost every time, it's because another mother-daughter book club has started as a result of my book. I'm awed by the momentum of mother-daughter book clubs. A grassroots movement that started with one club, now hundreds of mother-daughter book clubs are cropping up. In fact, the idea has taken off into all kinds of combinations. There's a mother-son book club and a father-son book club in Chicago. There's an older group of 16- and 17-year-old girls and their mothers who meet in Washington. There are several grandmother-granddaughter clubs, and one of my favorites, mentor-mentee clubs. These mentors are young volunteers, like Big Sisters, who come together to form support groups for girls in their own environments.

There are also school-based clubs, where not every parent may participate, but all the girls do. When I did a workshop for 80 Girl Scout troop leaders, every one of them left saying they were going to start clubs within the troops. A book club brings mothers into the troop, and it gives the troop leader one less activity to plan for the monthly meeting.

In talking to people all over the country, I've found that what they want to know most is how to make up questions for discussion. They want a tool, a resource to rely upon. So it seemed the next logical step was to provide that tool—a collection of good books and, more important, good discussion questions. *100 Books for Girls to Grow On* grew out of my desire to give fledgling book clubs a way to keep going. I wanted to make it easy for people who have book clubs to have great discussions.

I also wanted to encourage parents and children who are not in book clubs to read together, and to help broaden their reading experience. *100 Books for Girls to Grow On* helps you select books to read and provides insightful questions so that you can have a wonderful, satisfying one-on-one discussion with your child. From reading my summaries and in-depth questions, you can tell whether a book is right for your child today, much later on, or perhaps (rarely) not at all. After all, children have personalities. Books have personalities. The key is to find a great match, and I think this book can be your matchmaker.

When it comes to criteria for picking the books we read in my club, I'm off the hook: the girls choose! For *100 Books for Girls to Grow On*, I had to narrow down the enormous field of young adult literature. What an adventure! I read over 400 books and I discovered all over again how great books for this age group are.

Since I decided to write this second book, Morgan started recommending books to me, too. I read everything she suggested and most of the time I agreed with her. A lot of the choices in these pages are hers.

Although I applied certain rules and guidelines in my search, it mostly came down to selecting books that I thought were well written, held my interest, and would make for great discussions. (Yes, I do have favorites among them: *A Wrinkle in Time; The Golden Compass; The Eye, the Ear, and the Arm; Habibi; Walk Two Moons; Another Way to Dance; Dark Side of Nowhere.*) The subjects range from animals and relationships to history and science fiction. They aren't always about overtly mother-daughter issues. However, there are a few common motifs: coming-of-age stories; interesting, and often slightly offbeat, characters; and multidimensional plots with two or three stories going on at once. One of the things I've learned during this process is that a book doesn't have to be about mothers and daughters to raise "mother-daughter issues."

You'll notice a wide range of complexity within these selections, and that's quite deliberate. Everyone reads at a different level, so there are some books, like *The Hundred Dresses,* that are relatively easy yet still offer rich discussion opportunities, and some like *To Kill a Mockingbird,* that are more advanced. I've learned from our club that the length or difficulty of a book doesn't dictate how long or involved the discussion will be. One time we picked a short picture book—a 15-minute read—which prompted as much discussion as a full-length novel.

I chose certain older titles in order to encourage more bookstores to stock them and to keep them in print. One such title is *The Friends,* by Rosa Guy. It was really popular when it first came out in 1974, and the author has gotten a lot of acclaim outside this country, but it's not on everybody's shelf. Another one is *Ballet Shoes,* by Noel Streatfield. I tried to include worthy new books, as well. My trusted advisers at Cheshire Cat bookstore in Washington, who have helped me ever since we started our club, came to my aid once again in suggesting other new books. All the books I suggest can be ordered by your bookstore.

I eliminated certain traditional classics to avoid overlap with school reading assignments. For example, *Anne Frank: Diary of a Young Girl* is the classic example of its genre, but I've included several other diary books of which you may not be aware, such as *Red Scarf Girl: A Memoir of the Cultural Revolution,* by Ji-Li Jiang.

All the books have a strong, fully developed character—whether it's a boy or a girl—who can teach you something. And many different genres are represented. Some of the books are sheer fantasies. Several are historical fiction, which I don't normally read, but in my explorations I found a few I loved. For example, *The Poison Place: A Novel,* by Mary E. Lyons, was a nice way to learn something about African-American history, and the relationship between the daughter and the father reminded me of mine with my grandfather.

As an African-American, I chose lots of African-American-themed books—and many for other ethnic cultures. Obviously, race and culture play a huge role in our society today, but each of us experiences that differently. A book I really liked, *Habibi,* is about an Arab-American family that moves back to Israel. Even some of the science fiction talks about difference. In *The Eye, the Ear, and the Arm,* the three detectives are mutants as a result of chemical warfare. They look and act different.

How did I draw the line on content in choosing books? How did I gauge what girls this age can handle? I drew the line based on my own morals and values. When I started our own club, I picked up an interesting-looking book. It turned out to be about incest. I didn't think I would be comfortable talking about incest with my nine-year-old daughter, so I passed on that book. As the club progressed, we read a book that described physical abuse. It was hard, but ultimately rewarding. After that meeting, Morgan said, "If someone had come to me with that kind of a problem, I would have kept their secret. Now I would not keep somebody's secret. I would know that there are places that they can go and seek help." That made the difficult conversation worthwhile.

The most on-the-edge book included here is probably *Something Terrible Happened,* which addresses AIDS and interracial marriages. It's a wonderful book. I'm sure there are those who'd say AIDS and interracial marriages are inappropriate subjects to discuss, but I think we underestimate girls at this age. I think they can handle almost anything in a safe environment, which is what a book club or a parent can provide. It's so important to raise themes that encourage girls to tell stories and talk about experiences they don't share any other time. When we read *Something Terrible Happened* in the book club, one girl mentioned being at a mall with some of her white friends and running into a group of her black friends, who were speaking slang. She said she felt very awkward. It was the first time her mother had heard how she felt about going back and forth between different identities.

It's a fascinating, difficult world out there, and you have to listen to your kids. One of the hardest things for any adult, I think, is to let yourselves be open-ended enough to have a conversation or discussion take you wherever it wants to go. You can never tell where a book discussion will wind up. Pick a book because you like the book and the subject, not because you want to talk about a particular issue or find a prescription for an emotional ailment.

I gave the manuscript of *100 Books for Girls to Grow On* to another mother who showed it to her 11-year-old daughter one night. She had been reading *Tuck Everlasting,* so she turned to that entry and read the discussion questions. After a minute, her daughter looked up and asked, "Do you think it would be a blessing or a curse to live forever?" They then had a very fruitful discussion. *That's* my fondest hope for what this book might accomplish.

How to Use This Book

The entries in this book are arranged alphabetically by title, and each of them is composed of several different parts, which I'll describe below.

Summary

I begin with either a general description of the plot or my personal take on it. If I was especially moved by a book, or if it reminded me of a specific moment in the book club or with Morgan, I share that instead of—or in addition to—a straight synopsis.

Reading Time

I try to provide an average range of how much time it will take to read each book. I also note how many pages the book runs because everyone reads at a different pace. Although all the books should work for any girl between the ages of 9 and 13, some are skewed more toward younger girls, and others are for older readers. Whenever I feel this is the case, I note it in this section.

Themes

Ranging from friendship to jealousy, from loss to love, the topics covered in each book are briefly summarized to give you an at-a-glance sense of what to expect in discussions. Each theme is entered in the index as well, so you can find other books I've included that deal with similar topics.

Discussion Questions

This is the real focus of this book. The questions are designed to encourage heart-to-heart conversations while leading to the core of each book's characters and themes. They can't be answered with a simple yes or no, and there are no "right" answers. I try to offer enough questions for each book so that at least one or two will spark some interest. Mothers and daughters can ask these questions of each other, pose them at meetings, or simply read them to themselves. The goal is to get to know each other better, more than it is to educate.

You may want to modify or vary the questions as needed, to keep the discussion at a level the girls can manage. And if any of the questions seem likely to bring up an issue too close to home at the moment, you should obviously skip it.

About the Author

There is a short biography of every author in the book. I've found some of our best discussions have grown out of some tidbit of information the girls discovered about an author's background. If I include more than one book by the same author, I don't repeat the entry unless something specifically pertains to that title. (Authors who have multiple titles in this book include: Beverly Cleary, Sharon Creech, Nancy Farmer, Jean Craighead George, Virginia Hamilton, E. L. Konigsburg, Lois Lowry, Walter Dean Myers, Katherine Paterson, Jerry Spinelli, Mildred D. Taylor, and Laurence Yep.)

Beyond the Book...

This is where you'll find all kinds of book-related nonreading activities for you and your daughter to share. Some are obvious, like watching the movie version or preparing food a character describes. Other suggestions include field trips, crafts, games, and research. Keep it fun and keep it light. I'm sure many readers will always go right to this section, and I don't blame them!

If You Liked This Book, Try...

If the author has written other books, this is where some of them are listed. In addition, I point readers to other books that have similar themes or messages.

There is a quote I love from a *New York Times Book Review* article called "Fifty Years of Children's Books," that spells out why I want you to think of this book as a road map, not a bible.

> *The impossibility of issuing blanket recommendations is that in choosing books there is no set rule of thumb for this or that child which can be followed by parents as a cook follows a recipe, beating her eggs and thinking about something else. . . .*

Think about how you talk with your child—do your homework, clean your room, all the "do-it" questions. Or why did you do this, why didn't you do that? We don't sit down often enough and just have conversations. That's what I hope this book will help you to do. I guarantee that your kids are thinking about all these issues, even though you may not be talking about them. Even when you have the best intentions in the world, girls dread knowing a mother-daughter, heart-to-heart is coming. Morgan can't stand it. She'll say to me, "I don't want to hear another one of your little speeches." I'm not asking her what she thinks. I'm just telling her what I think.

Reading and discussing books together is a much more subtle, not to mention effective, way of guiding children rather than lecturing them. In these pages is a range of books that will open a door into your child's mind or life. No matter what issue you want to discuss, one of these 100 books is going to bring it out. I urge you to stop underestimating your children. Unless you talk to them in some kind of meaningful way, there's no outlet for them to hear your opinions, your values, and your morals. By discussing books with my daughter, I'm putting my spin on what she hears and reads and learns. Whether it's AIDS or eternal life—whatever the issue is—I'm commenting based on my own experiences. She knows where I stand.

I keep reminding people that books are fun. It's simple. Read with your child. That's what every parent should be doing to open up lines of communication. How you read together, when you read together, all the rest of it is just dressing. Read with your daughter—and watch her grow.

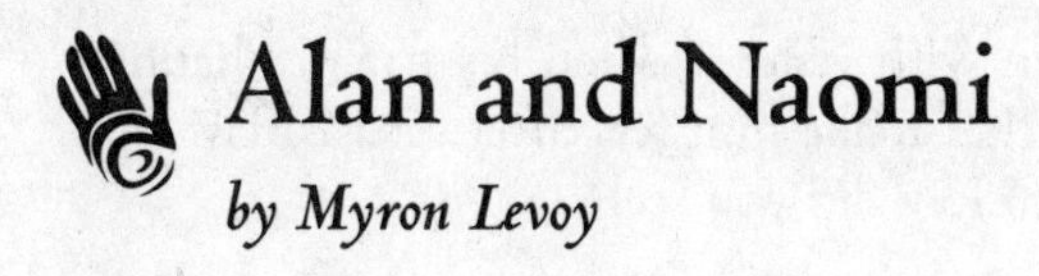

Alan and Naomi

by Myron Levoy

The specter of the Nazi Holocaust shadows the lives of two 12-year-olds living in New York City in 1944. Alan just wants life to be ordinary, yet his existence is changed forever when a young, Jewish French girl, traumatized by the war and the loss of her father, comes to live in his apartment building, and he reluctantly befriends her. This book is powerful, not only because it deals with the Holocaust, but because it portrays children learning to take responsibility for their own actions and facing the hard truth that things don't always work out.

READING TIME: 2–3 hours, about 192 pages
THEMES: prejudice, anti-Semitism, friendship, trust, betrayal, loss, guilt

Discussion Questions

- When Alan first sees Naomi, she reminds him of a lost puppy. Why does she strike him that way?
- When Alan is reluctant to be friendly to Naomi, his father tells him "In our life, sometimes when we're young, sometimes when we're old, in our life, once or twice, we're called upon to do something we can't do, that we don't want to do, that we won't do. But we do it." What does he mean? Can you think of any situations in your life when you were called upon to do something like that?
- Why do you think Naomi can communicate with Alan's ventriloquist dummy, and through her doll, but not with people?
- Naomi blames herself for her father's death at the hands of the Nazi Gestapo agents. How does that affect how she sees the world and herself?
- What does it mean to be a *mensch* to Alan's father? To Alan? What qualities or attributes do you think define a mensch?

- ✦ Alan is ashamed to be seen with Naomi when his friend Shaun is around. Why? How does that make him feel afterwards? Have you ever done that to a friend? How did you feel?
- ✦ How does being Naomi's friend change Alan? Why does it affect him that way?
- ✦ Why is it important that Alan acknowledges Naomi during one of his stickball games with his friends?
- ✦ How do Shaun and Alan misunderstand each other? What consequences does it have for their friendship?
- ✦ Why does Alan's fight with Joe Condello cause Naomi to run away? Despite Alan's best efforts, why can't Naomi recover from her war experiences? What do you think will ultimately happen to Naomi?
- ✦ Alan says that the Nazis got Naomi as surely as if they had thrown her onto a truck and taken her to a concentration camp. What does he mean? Do you think that's true?

ABOUT THE AUTHOR: Myron Levoy was born in New York City, and many of his stories portray the immigrant experience in the early part of this century. His books often depict characters who overcome adversity and whose struggles allow them to grow and become stronger. Alan and Naomi has been published in German and Dutch and in 1992 was made into a film.

Beyond the Book. . .

MAP: Look up Nazi resistance in an encylopedia or on the Web. Draw a map of Europe showing where the concentration camps were located, and where the Germans occupied different countries and regions. Read about the Warsaw Ghetto and its uprising.

HOLOCAUST MUSEUM: If you are from New York or Washington, DC, visit the Holocaust museum near you. If not, write to one and request information about the Holocaust and the exhibits they display. Because I live in Washington, I've had the opportunity to visit the Holocaust museum and found it tremendously moving. The curators have made an effort to personalize the Holocaust experience for patrons of the museum, making it much more real and intimate for people who don't have the Holocaust in their own histories.

MUSIC: Naomi responds to music and songs. Go to the library and find a songbook of World War II-era songs, or find a CD of popular tunes from that period, and play some of the songs.

MOVIE: Naomi enjoys a Marx Brothers movie; the physical comedy and slapstick humor transcend any language barrier. Rent one or two of the Marx Brothers' movies to watch together.

REFRESHMENTS OR FOOD MENTIONED IN THE BOOK: Serve chocolate bars and cherry soda during your discussion. Or make Alan's father's specialty, eggs scrambled with chopped mushrooms and onions. If you prefer, you can make a "miniature feast" like the one Alan and Naomi have on their picnic, with tiny triple-decker sandwiches with tomatoes and olives, little cakes with lemon, strawberry, and chocolate icing, and small bottles of milk.

IF YOU LIKED THIS BOOK, TRY...

Snow in August, by Pete Hamill—For older readers, it deals with similar themes of how someone survives the Holocaust and adapts to American anti-Semitism.

Anne Frank: Diary of a Young Girl, by Anne Frank—The diary of a young girl's life in hiding during the war, under Nazi occupation.

The Miracle Worker, by William Gibson—The story of Anne Sullivan's teaching of Helen Keller offers insight into how a dedicated and persistent individual can reach someone who, like Naomi, is locked into a world of her own—but for very different reasons.

Some Other Books by Myron Levoy:

The Witch of Fourth Street and Other Stories

The Hanukkah of Great-Uncle Otto

Alanna, the First Adventure

by Tamora Pierce

The first in a series of four books set in Medieval times, this novel tells the story of Alanna, an 11-year-old girl who dreams of becoming a knight. To do so means rejecting the female roles of her time and disguising herself as a boy. Alanna's cleverness, self-discipline, healing gifts, and determination set her on the path to fulfilling her dream.

READING TIME: 3–4 hours, about 216 pages
THEMES: gender roles, magic, identity, coming of age

Discussion Questions

- Why does Alanna want to be a boy? Have you ever gone through a period of wanting to be a boy? If so, describe why you may have felt this way.
- Within the medieval world in which Alanna is growing up, are there any ways that she can be a girl and still play a strong, honorable role in her society?
- Alanna and Thom have the close relationship of twins, yet they have very different talents, temperaments, and desires. Compare the twins.
- What does Alanna see as the expectations and limitations of being a girl?
- Discuss the upbringing of boys and girls during medieval times from information you learned from the book. How is it different than in today's world?
- Describe the training that knights received during medieval times. Why does this training appeal so much to Alanna?
- In many ways, the training of medieval knights is similar to the military training young men have always undergone to become sol-

diers. There are rigorous mental and physical challenges; some bullying and harassment to toughen the recruits; and an expectation of unquestioning obedience to authority. In today's world, some young women have chosen to undergo this kind of tough training in military schools that were once all male. Can you imagine yourself ever attending a military academy or school? Discuss what you think of these institutions, why they appeal to you or turn you off.

Alanna's spirit held my attention throughout this entertaining and inspiring book.

- Some of Alanna's ordeals make her question her plan to become a knight. At one point, her experience is described in this way: "She was still a girl masquerading as a boy, and sometimes she doubted that she would ever believe herself to be as good as the stupidest, clumsiest male." Have you ever had the experience of trying out challenges that mostly boys participate in and felt you couldn't be as good as the weakest boy in the group? What was the outcome?
- What do you think are male strengths and female strengths? Which do you believe society values more? Do you agree or disagree that both kinds of strengths are valuable in their own way?
- Describe Alanna's feelings about reaching puberty and developing a female body. Do you think her strong reaction to reaching puberty is typical or not?
- Mistress Cooper, who helps Alanna deal with menstruation, offers this advice: "Your place in life can always change, whether you have the Gift or not. But you cannot change what the gods have made you. The sooner you accept that, the happier you will be." What do you think of Mistress Cooper's advice? Does Alanna accept this advice by the end of the book?

ABOUT THE AUTHOR: Tamora Pierce has had a number of different jobs in her lifetime: martial arts movie reviewer, housemother in a group home, writer for a radio production company, literary agent's assistant, and investment banking secretary. She was only 11 when she started writing and has since written numerous books for children and young adults. Her husband Tim Liebe is a writer/filmmaker. They live in New York City with their parakeet and two cats.

Beyond the Book...

MEDIEVAL ART: Mothers and daughters may enjoy studying art books from the medieval period to see how young men and women are depicted. Encourage observation and discussion about which gender is shown in passive or adventurous situations.

KNIGHTHOOD: Research medieval traditions surrounding knights and knighthood, and present your findings to each other or the group. Hold a mock knighting ceremony according to the rites and rituals you have learned about. Then talk about what it was like to go through such a ceremony, even though it wasn't real.

REFRESHMENTS OR FOOD MENTIONED IN THE BOOK: For lunch serve bread and cheese, like Coram carries in his bag while traveling on the road. For dessert, have cherry tarts, like those that were discovered missing from the castle kitchens.

IF YOU LIKED THIS BOOK, TRY...

Catherine, Called Birdy, by Karen Cushman—This diary-style novel portrays the trials of a spirited teenage girl growing up in medieval times (see p. 29).

Life in a Medieval Village, by Joseph and Frances Gies—This nonfiction work describes social, family, and childhood life in medieval times.

Song of the Lioness Quartet, by Tamora Pierce—Three more novels in the series tell about the further adventures of Alanna, the female knight.

Some Other Books by Tamora Pierce:

Sandry's Book
Wild Magic

Another Way to Dance

by Martha Southgate

Another Way to Dance is the story of a young African-American girl's passion for classical ballet dancing. When she is admitted to New York City's School of American Ballet, 14-year-old Vicki Harris achieves one of her dreams, but she also begins to question her desire to become a classical dancer. Is she good enough? Must she give up parts of herself—her emerging racial identity, her appearance, her family ties—to attain a particular kind of artistic perfection?

READING TIME: 2–3 hours, about 180 pages
THEMES: cultural identity, body image, discipline, individuality, coming of age, divorce

Discussion Questions

- Vicki's passion for ballet is something her parents both support and question. What parts of Vicki's ballet training do her parents object to? How does Vicki react to their objections?
- Vicki gives up a great deal for ballet. What is she sacrificing to pursue her dream? Do you think Vicki makes the right decisions?
- How does the classical ballet environment affect Vicki's identity as a person of color?
- Vicki says: "One thing about ballet, you're never out of sight of yourself." How does Vicki feel about the girl she sees in the mirror? How often in a day do you check yourself in the mirror?
- Discuss what you think of Vicki's statement: "I really want to be part of all that perfection up there. If I work hard enough, I can be part of it. Maybe then I won't stick out so much." Why does she feel like she sticks out? Do you think working hard will make her feel like she belongs?
- In what ways does the dancing environment bring out Vicki's feelings about who she is?

- Why do you think Vicki has such a strong crush on "Misha" Baryshnikov, the Russian ballet star? How does Vicki change after she actually meets Baryshnikov?
- Vicki's date, Michael, brings her to church one Sunday. He tells Vicki that his mother says: "...I'll appreciate it when I get older." What kinds of things have your mothers or fathers said you will appreciate someday? Mothers: Were your own parents correct about your appreciating certain things once you grew up? Share those experiences.
- Vicki is stung by the comment another dancer makes about being chosen for the ballet school because of affirmative action. How does Vicki deal with this remark?
- Vicki holds her head up high when she learns she hasn't been selected to move onto the next level of training at the ballet school. Why do you think she is able to handle her disappointment with dignity? Discuss some of your own disappointments and how you have handled them.
- What has Vicki learned about herself by the end of the book?

This book really hit home for me because Morgan takes ballet, but I think anyone would be moved by the passion and dedication Vicki shows toward her dancing.

ABOUT THE AUTHOR: Though born and raised in Cleveland, author Martha Southgate now lives in New York City with her husband and son. *Another Way to Dance* is her first novel, and draws on her experiences studying ballet as a teenager. Southgate has worked as a journalist for over 11 years.

Beyond the Book...

DANCE: If you, or your daughter, has ever studied dance, talk about your experiences and maybe even share photos of yourselves dancing. Invite others who devote significant parts of their lives to other arts such as music, writing, and art to talk about what kinds of sacrifices are required to achieve the highest levels of training in the arts.

AUNTS: Vicki and her aunt have a more carefree relationship than Vicki and her mother. Girls often forge close, warm relationships with their aunts just as Vicki does. Invite aunts to the book discussion of *Another Way to Dance.* Encourage the aunts and nieces to talk about their relationships and how they differ from their mother-daughter relationships.

BALLET: Go to the ballet. If possible, stay after the show and talk to the performers about how they got started and what becoming a dancer was like for them. Or read a biography of a ballet dancer as a follow-up book.

MOVIE: Watch one of the ballet movies mentioned in the book: *Turning Point* or *White Nights.*

REFRESHMENTS OR FOOD MENTIONED IN THE BOOK: Vicki is a careful and conscious eater who deprives herself at the table more than she indulges herself. Organize a potluck supper of chicken, rice, and peas—from the book—as part of the book discussion. Focus on the full enjoyment of delicious, lovingly prepared food without worrying about calories! During the meal, talk about how worries about body images often interfere with the plain enjoyment of food.

IF YOU LIKED THIS BOOK, TRY...

Ballet Magic, by Nancy Robison—A gifted young dancer is the wrong size for ballet but dances to success anyway.

The Moves Make the Man, by Bruce Brooks—A gifted basketball player is the first black student at his high school. The themes of identity, skill, discipline, and friendship parallel those in *Another Way to Dance.*

The Soul Brothers and Sister Lou, by Kristen Hunter—A 14-year-old girl and her friends discover their talents and identity through soul music.

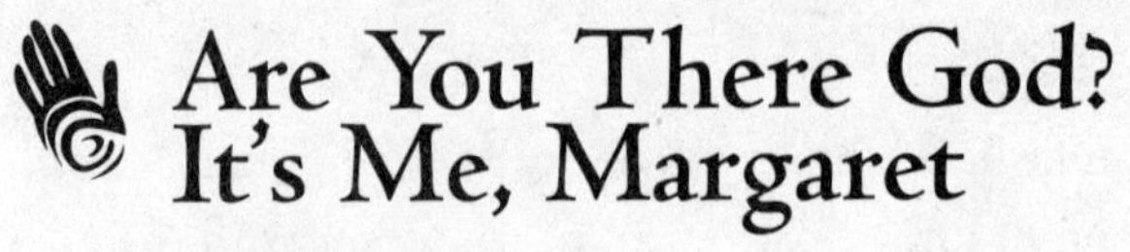

Are You There God? It's Me, Margaret

by Judy Blume

Going from 11 to 12 brings big changes to Margaret Simon's life. First there's an unwanted move to the suburbs. At a time when she's losing some of her sense of closeness to her parents, Margaret has to deal with new friends, teachers, religious identity, and puberty. Searching for someone to talk to, Margaret initiates a heartfelt dialogue with God to discuss this mixed-up, growing-up time of her life.

This book gives mothers a chance to share their own "firsts" with their daughters. Last year, Morgan and I attended our Girl Scout troop's mother-daughter tea, designed to educate girls about menstruation. The first exercise was to have mothers tell their daughters the story of how they got their periods. Thinking and talking about this reminded the moms what it was like to be where our daughters are now and gave the girls some insight into us, and what's going to happen to them.

READING TIME: 2 hours, about 150 pages
THEMES: moving, making friends, identity, religion, family, puberty, coming of age

Discussion Questions

- Do you have a "voice" you talk to about problems you can't discuss with anyone else like Margaret does? If so, where do you think that voice comes from, and how does it help you?
- Moving to a new place is full of challenges. What are some of the challenges Margaret Simon faces? Describe any similar challenges you faced if you've ever moved from one place to another.
- If you've ever changed schools, describe your first days at the new school or, if you haven't, describe the first day you started in mid-

dle school or junior high school. In what ways were those school situations like the first-day experiences Margaret has?

- ✦ Over the course of the book, Margaret succeeds in making several friends. Think back to friends you have made. How did you become friends? How long did it take before the friendship took off? How did you know when someone was really a friend?
- ✦ What are some of Margaret's feelings about her changing body? Are any of those feelings familiar to you?
- ✦ Why is it so important to Margaret not to get too far behind her friends in getting a bra and starting her period?
- ✦ Describe Margaret's relationship with her grandmother. How are relationships with grandparents, or aunts and uncles, different than those with parents?
- ✦ Margaret's grandparents have the following idea about religion: "'...A person doesn't choose religion...A person's born to it!'" Discuss what you think of this opinion.
- ✦ Do you think Margaret is a religious person even though she doesn't belong to any church? Do you consider yourself a religious person? Why or why not?
- ✦ Margaret samples several religions to find one to join. What do you think of the idea of a young person choosing her own religion?
- ✦ Margaret talks to God at least as much as she talks to her grandmother and parents. Since she is not formally religious, where do you think Margaret's ideas about God come from?
- ✦ In what ways is Margaret more grown-up at the end of the book than she was at the beginning? What do you think contributed to her growth?

ABOUT THE AUTHOR: Author Judy Blume has written 21 books for children and adults. Blume believes her own difficulties growing up helped her create characters that many young adults can identify with. She uses her books to let the young adults of today know that they are not alone, a thought that always comforted her. She has compared her favorite part of the writing process, rewriting, to a puzzle that she enjoys putting together. She currently lives in New

York City, not far from Elizabeth, NJ, where she was born. Her husband George Cooper, a nonfiction writer, thinks his wife is lucky because she gets to make things up, while she envies his ability to research and discover stories. They have three grown children and one grandchild, whose first word was, appropriately, "book."

Beyond the Book...

FIRSTS: Share some of the teen and preteen "firsts" you feel comfortable talking about: getting a bra; using deodorant; learning to dance with a boy; kissing a boy; going to a preteen party; and starting to menstruate. Point out similarities between your "firsts" and those of Margaret Simon. This sharing of preteen memories will help your daughter feel that such experiences are universal and normal.

PRAYER: If you are a person who prays, share any odd requests you've asked for in a prayer. Talk about the different ways prayer is practiced in various religions. If you already pray regularly, try experimenting with a new kind of prayer—if you don't, you might want to try it once and see what you get out of it.

REFRESHMENTS OR FOOD MENTIONED IN THE BOOK: Margaret's grandmother brings some deli food from the city to the "wilds" of suburban New Jersey. You might enjoy sharing a similar meal as part of the book discussion: hot dogs, potato salad, cole slaw, corned beef, rye bread, and pickles.

IF YOU LIKED THIS BOOK, TRY...

Getting Your Period: A Book About Menstruation, by Jean Marzollo or *It's a Girl Thing,* by Mavis Jukes—These books discuss both the emotional and physical changes girls experience during puberty.

My Friends' Beliefs: A Young Reader's Guide to World Religions, by Hiley H. Ward—This very readable nonfiction book examines the history, customs, and basic tenets of both major and lesser-known religions.

Some Other Books by Judy Blume:

Freckle Juice
Blubber
Otherwise Known as Sheila the Great

Ballet Shoes

by Noel Streatfield

The theater brings out latent talent in three unrelated orphans who belong to a somewhat unconventional family, and helps them discover their distinctive destinies in this engaging story. Though this book is over 50 years old, its message—that everyone has a gift but it's up to the individual to put it to good use—still rings true.

READING TIME: 2–3 hours, about 235 pages. Some of the "Britishisms", like pantomimes, jumpers, and crackers, may need explanations or translations.
THEMES: identity, family, self-esteem, ambition, selfishness, opportunity, competition, economic status

Discussion Questions

✦ What kind of family were Paulina, Petrova, and Posy brought up in? How did they feel about that? Imagine how you might feel if your family were a "collection" of children from very different backgrounds. What would make that group of people feel like a family?

✦ What kind of relationship did the girls have with Sylvia, a.k.a. "Garnie"? How is she like a parent to them, and how is she different?

✦ How do the boarders who come to live in the family's house have an impact on the children's lives? What do each of the girls learn from them that will affect their future?

✦ Why is there a special bond between the oldest girls, Paulina and Petrova? How do they feel about Posy? Why?

✦ Petrova's talents lie in different directions from her sisters'. How does that make her feel when she is sent to study at the Academy with them? Have you ever been put in a situation where you've felt that way? What was it like?

✦ When Posy, the youngest, is selected to study privately with

Madame, the head of the Academy, how do Paulina and Petrova feel? How would you feel if a sibling were singled out for special treatment?

- When Petrova makes a special friend out of the boarder, John Simpson, who buys a garage, how does that change the way she feels about herself? When her secret is revealed, how does it change how her sisters feel about her?
- How does Paulina change when she gets a major acting part as Alice? What happens to her relationships with her family and friends as a result?
- What kind of person is Posy? Would you want to have her as a sister? Would you want to have her as a friend?
- Which Fossil girl do you like best? Why?
- What do you think happens to the girls when they grow up? If you were going to write a story about the Fossils as adults, what would you create?

ABOUT THE AUTHOR: Noel Streatfield was born in England in the 1890s and began her career as a dancer and Shakespearean actress. She started writing in 1930 when she became a book critic for *Elizabethan* magazine. She is best known for her books about ballet, inspired by her own experiences. Her novel *Ballet Shoes* was made into a television series in 1976. Her hobby was collecting wildflowers. Before her death in 1986, she wrote over 75 books, including a series under the pseudonym Susan Scarlett.

Beyond the Book. . .

BALLET: Attend a ballet performance by children in your community, or watch a show performed by a children's theater group. Do some research on what it takes to become a ballet dancer, and the kind of training young girls go through. Maybe a local dance school would provide a one-time lesson so you could see what ballet is like.

MOVIE: See if you can find a Shirley Temple film called *The Blue Bird*, which is one of the plays that Paulina and Petrova perform, or rent one of the many versions of *A Midsummer Night's Dream.*

COSTUME: Design and make your own costume for your own performance of either *A Midsummer Night's Dream* or *The Blue Bird.*

REFRESHMENTS OR FOOD MENTIONED IN THE BOOK: You may want to serve some of these: tea, cake, milk and cookies, ice cream sodas, hot chocolate with whipped cream.

IF YOU LIKED THIS BOOK, TRY...

Noel Streatfield wrote many other books, which unfortunately are somewhat hard to find as many are no longer in print. See if your school or local library has a copy of *A Painted Garden,* which features some grown-up Fossils.

Some Other Books by Noel Streatfield:

Theater Shoes

Dancing Shoes

A Bone from a Dry Sea

by Peter Dickinson

Vinny accompanies her estranged father, an archaeologist, on a dig in Africa, where she helps make an amazing discovery: an artifact made from a dolphin bone *four million years ago.* In alternating chapters, we go back four million years to a marsh where the prehistoric child Li lives with her tribe. At the end of Li's story, she is using a pointed rock to make a dolphin bone into a hair ornament! Li, a creative thinker herself, is the proof of a revolutionary theory about human evolution—but no one can ever know her story.

This is the only young adult book I have come across that alternates characters and story lines between chapters. Peter Dickinson does a superb job weaving two incredible stories into one fascinating novel.

READING TIME: 2–3 hours, about 199 pages
THEMES: family, divorce, science, evolution, creativity, courage

Discussion Questions

- Vinny's father explains how little we know about how humans evolved from four-legged, apelike creatures "something like us...walking on two legs." He also points out that the excavation is taking place in an area that was once ocean. How do these facts explain why the story of Li and her people is related to Vinny's father's work?

- Vinny has read about the "sea-ape theory" in the works of Elaine Morgan. She finds the theory exciting, and she even finds evidence to support it; yet her father becomes angered when she voices her ideas regarding this theory. What in the story of Li, although unknown to the modern characters in the book, actually proves Morgan's (and Vinny's) theory to be valid?

- Why does Vinny's father get angry at her for discussing the sea-ape theory? Do you think Vinny should have dropped the subject, or would you, too, have kept bringing it up, even if it caused conflict?

Why? Do you think the tensions between scientists and nonscientists on the topic of evolution will ever be resolved?

- ✦ Vinny's parents did not get along well. From the descriptions you've read of each of them, do you think that it would have been better for Vinny if they had stayed together? Why do you think their marriage broke up?
- ✦ What do you think of the way Vinny handles her parents? Do you think it was fair how her parents dealt with her concerning their relationship?
- ✦ Even if your parents are not divorced, as are Vinny's, you may have experienced times when your parents disagreed. Do you think children should get involved and take the side of either their mother or father? How does that situation make you feel?
- ✦ Vinny and May Anna get along very well. In real life, a child sometimes does get along with a single parent's girlfriend or boyfriend, but often these relationships can be difficult. What might cause these difficulties? What circumstances in the story might make it easy for Vinny and May Anna to get along?
- ✦ In which ways are Vinny and Li alike? In which ways are modern humans similar to Li's people?
- ✦ Moms: What did you think about Vinny's father's boss? Have you ever had to work for someone who was difficult? Would you have handled the situation any differently? What advice would you give your daughter about dealing with a boss?

ABOUT THE AUTHOR: Peter Dickinson received his Bachelor of Arts degree in 1951 from King's College, Cambridge, quite a distance from his home in Livingstone, Northern Rhodesia (now Zambia). Some of his careers have included military service in the British army, editor of a well-known magazine, and author. Though he has written many books for young adults, he has also written novels for adults, and a television series for young people. He has had great success in his writing career and has won numerous awards. Dickinson currently lives in London.

Beyond the Book...

REVOLUTIONARY THINKERS: Use the library, a set of encyclopedias, or the Internet to find information on the following revolutionary thinkers whose ideas were rejected at first: Columbus, Galileo, Roger Bacon, William Harvey. You might also learn more about Charles Darwin's *The Origin of Species,* which caused a revolution in natural science.

MUSEUM: Visit a museum of natural history where you can examine examples of fossils and learn more about what we find out from fossils, how they are excavated, and how they are identified and dated.

FOSSILS: If at all possible, go on a local fossil dig (sometimes sponsored by local museums or nature institutions). Museum shops and nature stores often sell fossil excavation kits. Buy a kit (or put one together), and do some research about where in your area you would be most likely to find fossils. Go there and dig—who knows what you'll find!

SCREENPLAY: Choose one of the "Then" chapters about Li and write it up as a screenplay. You will have very little dialogue, so most of your writing will be describing the setting and the appearance, actions, and expressions of the characters. Choose appropriate music for the background. Then be the director as others act out the parts.

IF YOU LIKED THIS BOOK, TRY...

Bring to a Boil and Separate, by Hadley Irwin—This book chronicles a younger girl's experience with divorce and how she copes with it during her traumatic 13th summer.

How Did We Find Out about Our Human Roots? by Isaac Asimov—This is an overview of human evolution in theory and fossil finds.

Digging the Past: Archaeology in Your Own Backyard, by Bruce Porell—This is a sound introduction to the science of archaeology and its methods of investigation.

Some Other Books by Peter Dickinson:

Chuck and Danielle

The Gift

The Lion Tamer's Daughter and Other Stories

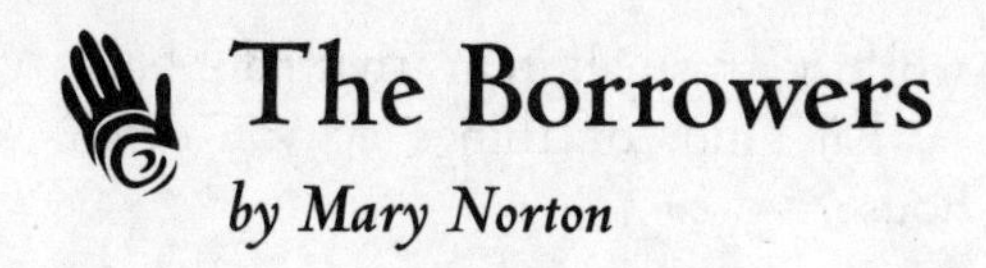

The Borrowers

by Mary Norton

In this whimsical tale, the miniature Clock family, known as the Borrowers, have set up housekeeping under the floorboards of an old house. Rarely seen and living in secret, these little people borrow loose items from the "human beans" upstairs. Their cozy hideaway is exposed, and their lives endangered, when their restless daughter makes contact with a nine-year-old boy who discovers them. The story of their encounter is woven with charming details and plenty of adventure.

READING TIME: 2–3 hours, about 200 pages
THEMES: humanity, independence, tolerance

Discussion Questions

- Each of the Borrowers has very human traits. Describe Homily's human qualities. What about Pod's?
- In what ways is Arrietty very much like an adolescent "human bean?"
- What does Arrietty think when she sees the larger outside world? What does Arrietty's first taste of freedom feel like to her?
- What observations do the Borrowers and the boy have to make about each other? If your family was being observed at close range by secret inhabitants, what do you think they would make of you? Why?
- Imagine there are huge creatures living on earth and human beings are the small ones who can't be seen. How would that feel?
- Arrietty and the human boy have a conversation about the difference between borrowing and stealing. What does Arrietty say the difference is? Do you agree?
- The boy offers the Borrowers a wealth of objects from his world. How does he feel about the tiny family he discovers living right underneath him? Why is he so upset by the reaction of the human adults when the Borrowers' existence is discovered?

✦ The Borrowers furnish their world with small items that have been lost in a human household. What kinds of things do you often find missing from your own house?

ABOUT THE AUTHOR: Mary Norton was born and lived in England, where she worked as an actress, playwright, and author. She has said many of the ideas for her books grew out of things she remembered imagining as a child. This is the case with *The Borrowers,* which was drawn from a childhood fantasy she had about tiny people living just out of sight of humans. Norton died in 1992.

Beyond the Book...

FOUND OBJECTS: Prior to the book discussion, create new items from old found objects, small enough to be used by the Borrowers. Mothers and daughters might particularly enjoy re-creating the very items mentioned in the book. While working on this project, discuss why many people find the idea of miniature worlds so appealing.

DIORAMA: Make a diorama of a space where the Borrowers could have lived, based on a spot in your own home. Collect small objects and hypothesize about what a small family could use them for.

MOVIE: See the British Broadcasting Corporation's movie version of *The Borrowers* or the 1996 movie *Honey, I Shrunk the Kids.*

REFRESHMENTS OR FOOD MENTIONED IN THE BOOK: Prepare and serve the kind of traditional English tea the Clock family so cherishes. Research what foods constitute a proper English tea.

IF YOU LIKED THIS BOOK, TRY...

Bedknobs and Broomsticks, by Mary Norton—Three children accompany a student witch on her zany adventures.

The Borrowers Afield; The Borrowers Afloat; The Borrowers Aloft; The Borrowers Avenged, by Mary Norton—The Clocks have further adventures.

The BFG, by Roald Dahl—A friendly giant delivers good dreams to children.

Some Other Books by Mary Norton:

Are All the Giants Dead?

Bridge to Terabithia

by Katherine Paterson

Jess and Leslie, two lonely children, form an unlikely but fast friendship in this bittersweet novel. They create an imaginary land that offers a needed escape from the obstacles of their daily lives. Together they reign as king and queen, until one tragic morning. Jess must come to terms with this tragedy before he can finally understand all he has learned from Leslie.

I could really relate to Jess's experience in this book. Almost a year ago I lost one of my best friends when she died suddenly of an aneurysm. We were the same age, our husbands and sons were best friends, we both loved art and jewelry, and volunteered for some of the same community organizations. For us, Terabithia was her house—it was where we gathered for parties and private chats. When she died, my first reaction, like Jess's, was anger. Thanks to previous book club discussions, though, I had the words to talk about her death with my family and friends, and moved past the anger to mourn for my good friend.

READING TIME: 1–2 hours, about 128 pages. Some children may be disturbed by what happens to Leslie.
THEMES: individuality, stereotypes, gender roles, imagination, privacy, death, loss, grief

Discussion Questions

- ✦ Why is running so important to Jess? Do you have an activity you feel passionately about? What is it, and why does it matter so much?

- ✦ Jess's mother appears to treat him quite differently than she treats his four sisters. How does that make Jess feel about himself? How does that make him feel about his mother? Have your parents done things that triggered similar feelings in you?

- ✦ Meeting Leslie seems to change Jess's feelings about running. Why? How does that influence his feelings toward her?
- ✦ How does Jess feel about being labeled the "crazy little kid that draws all the time"? How do you feel if people label you for a particular talent or skill you have, or because of your appearance or how you dress?
- ✦ How does Miss Edmunds respond to Jess's art? Why is that important? Have you ever had a teacher who took a strong interest in you?
- ✦ Leslie and Jess have many similarities, but they're also pretty different. Describe, specifically, what traits they do or do not have in common. What specific qualities does Leslie have that Jess admires, envies, or wishes he had? How can you tell?
- ✦ Jess observes about Leslie, "Lord, the girl had no notion of what you did and didn't do." What is he referring to? Does he see that as a good or bad thing? How does the rest of the community see it?
- ✦ Why is Leslie unhappy? Does she feel comfortable with the other kids at school, besides Jess? Why or why not?
- ✦ How are Leslie's parents different from other parents? How does that make Leslie feel? How would it make you feel to have parents like hers?
- ✦ Why do Jess and Leslie create Terabithia? What does Terabithia represent for them? Have you ever had a secret place like that?
- ✦ Jess feels that "Leslie was more than his friend. She was his other, more exciting self—his way to Terabithia and all the worlds beyond." What does this mean? Discuss.
- ✦ Have you ever had that kind of relationship with a special friend?
- ✦ What's significant about the Christmas presents—the puppy and the art supplies—that Jess and Leslie exchange?
- ✦ Jess and Leslie both have difficult relationships with their mothers, but for different reasons. Why is Jess's relationship with his mother so hard? What about Leslie? Why don't she and her mother get along?

- At first, Jess's sister Brenda thinks he doesn't care that Leslie has died. Why does she think that?
- Why was Jess angry with Leslie when he first hears about her death? Why doesn't he want to be at the funeral? What makes him stop being angry with her for dying?
- After Leslie drowns, why do you think Jess throws her present into the gully? How does Jess feel about Leslie dying? Have you ever felt that way about someone close to you who died?
- What did Leslie's friendship teach Jess? The book says, "It was up to him to pay back to the world in beauty and caring what Leslie had loaned him in vision and strength." What does that mean?
- How do you think Jess will live his life differently because of his friendship with Leslie? What do you think Jess will do with his life?

ABOUT THE AUTHOR: Katherine Paterson firmly believes that, as an author, you can't wait for inspiration to hit before you begin writing. It's a job like any other, she says. But *Bridge to Terabithia* did grow out of Ms. Paterson's real-life experience. Her son, David, lost his best friend, eight-year-old Lisa Hill, when she was struck and killed by a bolt of lightning. Ms. Paterson says she wrote the book "to try to make sense out of a tragedy that seemed senseless." She based the characters on a community in rural Virginia where she taught while training to be a missionary. In fact, a teacher she once met at a meeting in Virginia told Ms. Paterson that when she read *Bridge to Terabithia* to her class, she found out that one of the students' mothers had been in Ms. Paterson's sixth-grade class. She's glad to know that these children are now grown up with children of their own and know about the book. She hopes that they can tell by the book how much they meant to her.

Beyond the Book...

ART MUSEUM: Art is especially important to Jess, who feels privileged when his teacher, Miss Edmunds, takes him to the art museum in Washington. Take a trip to a local art museum and discuss what pieces of art are particularly moving or meaningful, and why. If possible, go on the Internet and explore the National Museum galleries online.

TERABITHIA: Using papier-mache, construction paper, and other materials, create your own version of Terabithia, and discuss what elements are important in the invention of a hideaway.

REFRESHMENTS OR FOOD MENTIONED IN THE BOOK: Serve foods that Jess and Leslie enjoyed in their secret spot, like dried fruit, crackers, and peanut butter. Talk about what people would choose to eat if they could escape to a secluded refuge, and why they would make those selections.

MOVIE: Rent the movie *My Girl,* another bittersweet story of a friendship that ends in tragedy.

IF YOU LIKED THIS BOOK, TRY...

The Secret Garden, by Frances Hodgson Burnett—Lonely and misunderstood children find sanctuary in a secret place that they create.

Some Other Books by Katherine Paterson:

Jacob Have I Loved (see p. 153)

The Great Gilly Hopkins (see p. 113)

Lyddie

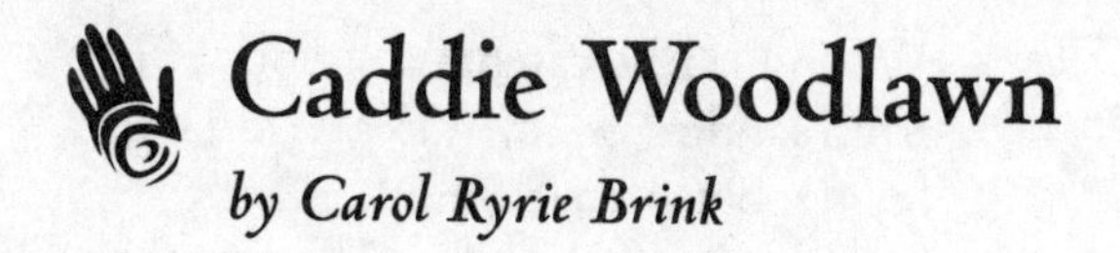

Caddie Woodlawn

by Carol Ryrie Brink

For 11-year-old Caddie, pioneer life in Wisconsin in 1864 is a series of daily adventures that most girls of her time didn't experience. At her father's insistence, Caddie is being raised like her brothers rather than like most girls of the period. This means plenty of hard work, outdoor life, physical challenges, and, above all, the freedom to be her spirited, adventurous self.

READING TIME: 3–4 hours, about 265 pages
THEMES: family, responsibility, gender roles, cultural identity, coming of age

Discussion Questions

- ✦ Describe your first impressions of Caddie Woodlawn. In what ways do you identify with her?
- ✦ Mr. and Mrs. Woodlawn have strong, but different roles to play in the household. Describe their individual ways of parenting Caddie. Which parent seems to have more influence on Caddie?
- ✦ The author says: "When Father or Mother made a decision, the Woodlawn children accepted it as final. There was very little teasing for favors in a large pioneer family." Why do you suppose it was important for pioneer families to operate in this way? How are big decisions made in your own family? In what way is your own family similar to or different from the Woodlawns when it comes to family decision-making?
- ✦ Why is Caddie allowed a different kind of upbringing than most girls of her time?
- ✦ Compare Caddie's household chores to your own.
- ✦ What are Caddie's feelings about someone "making a young lady out of her?" Describe what that "young lady" training was back in 1864. What kinds of "young lady" training exist today?

Caddie is so contemporary in her thinking that on several occasions I forgot that the book was set in the 1800s. She's a great role model for girls today.

- When Caddie catches a serious cold from landing in icy water, Caddie's loving, but puzzled mother, asks: "...why can't you behave like a young lady?" Why can't Caddie behave like a young lady of that period—or why doesn't she want to?
- Compare a typical day at your own school with Caddie's morning at her one-room schoolhouse. In what way are the students of that time similar to or different from students you know?
- How does the Woodlawns' relationship with the Indians differ from that of their pioneer neighbors?
- Describe how Caddie prevents a massacre from taking place between the pioneers and Indians.
- Caddie forms a quick impression of her cousin, Annabelle. How and why does that impression change for Caddie?
- After her cruel practical joke at Annabelle's expense, Caddie feels a deep shame that changes her. Afterwards, "...something strange had happened to Caddie in the night. When she awoke she knew that she need not be afraid of growing up." What has she learned that helps her face growing up? Is growing up the same as becoming a lady?
- After her mother punished Caddie for her practical joke, Caddie's father tries to comfort her by saying that "somehow we expect more of girls than of boys." Why did he say this? How is this statement true in the novel? Do you believe it is true in your life?
- How does Caddie's father define "a woman's work"? By the end of the novel, do you think Caddie has "run with the colts long enough"? What does that mean?

ABOUT THE AUTHOR: Carol Ryrie Brink, an accomplished short-story writer and novelist, lives in St. Paul, MN, with her husband and two children. *Caddie Woodlawn* is based almost entirely on the childhood experiences of Ms. Brink's grandmother and the stories she told Carol in her youth. Ms. Brink has compared writing *Caddie Woodlawn* to "weaving a tapestry" as she pulled together the threads of her grandmother's memories. *Caddie Woodlawn* is her second novel and won the Newbery Medal.

Beyond the Book...

MAP: On a United States map, trace the journey Caddie's mother made from the bustling city of Boston all the way to the Wisconsin frontier. Imagine what that journey must have entailed: saying goodbye to family members, perhaps forever; trading urban life conveniences for unknown challenges. Who would have liked to experience such challenges?

GENERATIONS: Travel back one or two generations by interviewing the oldest people you know. Ask about their earliest childhood memories and the stories *they* remember being told by their own parents. If possible, invite some of them to the book discussion of *Caddie Woodlawn.*

NICKNAMES: Come up with a word or name that describes the essence of one of your friends. Discuss the results and the rationale. With your mother, think of names for siblings and fathers.

GIRLS' ROLES: Get some old etiquette and parenting books from the library to see how children, particularly girls, were raised during different time periods. Consider how expectations of girls' roles limited or liberated them at different times in history. Are there any old customs today's girls wish were still promoted?

REFRESHMENTS OR FOOD MENTIONED IN THE BOOK: You may enjoy baking a loaf of bread from scratch to share at the book discussion. Remember that bread-baking was a constant, time-consuming task of everyday life and central to the life of Caddie Woodlawn. Consider whether this activity brought mothers and daughters closer together every day. Discuss or suggest everyday activities that might promote closeness between parents and children.

IF YOU LIKED THIS BOOK, TRY...

Addie Across the Prairie, by Laurie Lawlor—A pioneer family travels through Indian country on their way across the prairie.

Little House on the Prairie, a series by Laura Ingalls Wilder—Pioneer family experiences life on the frontier (see p. 170).

One Day on the Prairie, Jean Craighead George—A nonfiction book about a day the author spent at the Prairie Wildlife Refuge in Oklahoma.

Some Other Books by Carol Ryrie Brink:

Baby Island

A Chain of Hands

Snow in the River

Catherine, Called Birdy

by Karen Cushman

Through her diary, a fictional 14-year-old girl from the year 1290 speaks to contemporary girls about her place in medieval society. Catherine, nicknamed Birdy after the birds she tends, uses this journal to express the thoughts, emotions, and opinions she is forbidden to voice publicly or within her own family. Catherine's inner turmoil over her prescribed role in medieval society is expressed with wit, spirit, and her indomitable belief in her own inner worth.

The intimate experience of reading someone's diary reminded me of the journal I kept when I was young, and the time my mother discovered and read it. We'd had an argument and I wrote that I "hated" her. I brought this episode up at a club discussion of diaries and we had a good talk about the different experiences group members had with diaries and journals.

READING TIME: 3–4 hours, about 217 pages
THEMES: duty, gender roles, independence, religion, coming of age, identity

Discussion Questions

✦ What is your first impression of Catherine when you begin to read her diary?

✦ Why does Catherine keep a diary? How does she feel about writing in it? If you have ever kept a diary, discuss what purpose it served in your own life.

✦ Describe some of the restrictions and expectations Catherine is expected to follow as a young girl in medieval society?

✦ What specific activities would Catherine prefer to be doing instead of learning to be a "lady in training"?

✦ Catherine says of herself: "I am stubborn, peevish, and as prickly as a thistle." Do you think these are faults or virtues in Catherine's character?

✦ Describe Catherine's relationships with her mother, her father, and her nurse, Morwenna.

✦ When Catherine expresses to her mother her longing to be a song maker, her exasperated mother replies: "Song maker, Birdy? Don't stretch your legs longer than your stockings or your toes will stick out. . . . You are so much already, Little Bird. Why not cease your fearful pounding against the bars of your cage and be content?" Why can't Catherine be content with her life?

✦ Catherine makes a long list of things medieval girls are not allowed to do—everything from laughing too loudly to ice skating. Do you think there are things today's girls are discouraged from doing? Discuss what today's restrictions might be.

✦ Catherine describes a day of "lady lessons" and says: "It is impossible to do all and be all a lady must be and not tie oneself in a knot." Discuss some of the conflicting social expectations of today's girls that sometimes make them feel tied in knots.

✦ Catherine dates each of her diary entries with a saint day, some of which are quite funny. Share some of your favorite saint day descriptions from Catherine's diary.

✦ Why do you think Catherine chooses to raise birds? What does her favorite choice of a pet say about her? Why does she set her birds free at the end of the book?

✦ In response to Catherine's disguises and attempts to escape her lot in life, a traveling woman suggests that she may someday be asked: "Why were you not Catherine?" In what ways does Catherine become who she really is at the end of the book, despite stepping into some of the roles her family and her society expect of her?

ABOUT THE AUTHOR: Karen Cushman is a storyteller, but until her husband asked her to write down her stories, she didn't consider herself a writer. Though she began writing late in life, author Karen Cushman has had great success. Her first two novels won the Newbery Medal, as well as many other honors. Her interest in how ordinary people lived in other times led her to research and write her historically centered books, which focus on everyday people instead of kings, generals, and other characters common in historic tales. She is an assistant director of the Museum Studies Department at John F. Kennedy University in the San Francisco Bay area. Karen Cushman and her husband now live in Oakland, CA, with their daughter Leah, and their many pets.

Beyond the Book...

NEEDLEWORK: While Catherine hated needlework because it was demanded of her, this relaxing activity has often brought women and girls together. Book club members who do some kind of needlework, such as knitting, quilting, sewing, embroidering, might enjoy teaching some simple stitches to the nonsewers in the group. Perhaps the group can complete a simple round-robin project as they discuss *Catherine, Called Birdy.*

JOURNAL: Reading this book reminded me of a discussion our group once had about diary and journal keeping. Several of our members are diligent about keeping a daily diary while others—myself included—have little or no interest in it. Talk about journals and diaries—what they are good for, why people keep them, what the difference is between the two. Have each member start keeping one for a week or a month—if they don't already—and talk about the experience at the next meeting.

ART HISTORY: Look at some of the historical picture books about medieval times listed in *IF YOU LIKED THIS BOOK, TRY...* with the discussion group, or take a trip to a museum with some medieval works of art to learn more about the religious and daily life of that period.

REFRESHMENTS OR FOOD MENTIONED IN THE BOOK: Nuts, cheese, and apples are the most readily obtainable—and familiar—foods mentioned in *Catherine, Called Birdy.* However, many foods less common today are mentioned as well—everything from pigs' stomachs to eels.

Everyone might enjoy a discussion of the foods and mealtimes Birdy details so hilariously in her diary. Try to find one food available today that no one has ever had and prepare it for the next meeting.

IF YOU LIKED THIS BOOK, TRY…

Alanna: The First Adventure, by Tamora Pierce—A young girl, who wishes to become a knight in the Middle Ages, switches places with her brother (see p. 4).

Life in a Medieval Village, by Joseph and Frances Gies—A nonfiction work describing the details of daily village, family, and childhood life during medieval times.

Otto of the Silver Hand, by Howard Pyle—The son of a robber baron in medieval times is kidnapped, and his adventures lead him to a monastery.

Some Other Books by Karen Cushman:

The Ballad of Lucy Whipple
The Mid-Wife's Apprentice

Charlie Pippin

by Candy Dawson Boyd

Chartreuse "Charlie" Pippin can't seem to keep out of trouble. First, she can't understand why the principal doesn't approve of the various business enterprises she runs so successfully at school. And at home, everything she says and does seems to irritate her father, a Vietnam veteran whose emotional wounds from the war have never healed. It is Charlie's report on the Vietnam War and her visit to the Vietnam Memorial in Washington, DC, that finally help her and her father understand and accept one another.

When our club did this book, I was surprised by how little the girls knew about the Vietnam War—when our host took out a globe and asked the girls to locate Vietnam, it took them more than three tries! This book turned out to be a really "cool" way to learn history.

READING TIME: 2–3 hours, about 182 pages
THEMES: honesty, responsibility, parent-child conflicts, civil disobedience, protest movements, divorce

Discussion Questions

- ✦ When Charlie interviews people for her report on the Vietnam War, her opening line is, "Tell me what you know about the Vietnam War." How do most of her classmates respond? How would you answer Charlie? What have you read, heard, or learned about the Vietnam War and the protest movement against it?
- ✦ Discuss what you know about the war. Do you agree with Oscar or Ben? Have you formed an opinion about the Vietnam War yourself?
- ✦ Why does Mr. Rocker, the principal, forbid Charlie from selling things in school? Do you think he's right in doing this? How would Charlie have argued in favor of her business? What arguments could Mr. Rocker have used against Charlie's?

- ✦ If you were Charlie's mother or father, how would you react to Mr. Rocker's telephone calls? How would you deal with Charlie?
- ✦ What is the relationship between Charlie and her grandmother? Why do you think kids can often share their feelings with grandparents that they can't with their parents?
- ✦ When Charlie first brings up her project on the Vietnam War at home, she learns of a side of her parents she never knew—the dreamers. Her father says, "...giving up dreams for responsibility is life." What does that mean? What dreams did he give up?
- ✦ Mothers: Did you, like Charlie's parents, have dreams that you were never able to fulfill? If you feel comfortable talking about these dreams, share them. How do you feel now about those dreams? Daughters: What are some of your dreams for the future? Do you have a plan for fulfilling them?
- ✦ Granddad asks Charlie, "When you figure out how to make someone love you just the way you need them to, would you please come and tell me?" What does Granddad mean? What is Charlie looking for? What does he think Charlie has to learn about love and the way people love each other?
- ✦ How do you define the word "love"? Does everyone define love in the same way?
- ✦ Do you think Charlie was right or wrong to go to Washington with Uncle Ben? Give reasons for your opinion. Were lying to her family or giving up the trip her only options? Can you think of a way she could have made the trip without lying to her family?
- ✦ Katie Rose's parents are divorced, and she moves back and forth between her parents' separate homes. Daughters: Do you think this is the best arrangement for kids whose parents live separately? Which problems does this arrangement solve? Which problems does it create? Can you think of a better arrangement? Moms: If you are divorced, talk about what it was like for you when you were dealing with custody issues, and how you reached a settlement.

ABOUT THE AUTHOR: Candy Dawson Boyd was born in Chicago in 1946 and grew up listening to her African-American family's stories of hardships overcome. She became active in the Civil Rights movement while attending Northeastern Illinois State University, quit school to work with Dr. Martin Luther King Jr., and became a teacher in the early 1970s. She decided to write children's books as a result of her frustration over the lack of good reading material for her students. Boyd is an actress and a singer as well as an award-winning novelist.

Beyond the Book. . .

DOING BUSINESS: Like Charlie, try becoming an entrepreneur (but not at school!). Think of a need you could fill in your community. Then plan how to fill that need and how to sell your goods or services. You may or may not want to put your plan into action, but if you do, be sure to discuss your experiences and results.

PLAY REPORTER: Choose a subject about which you'd like to know more and write a report about it. Like Charlie, make interviewing one of your research tools. Try to choose a current controversial topic that lends itself to interviewing. Ask other people what they know about your topic and what their opinions are about it. You may want to return to the book and analyze some of Charlie's interviewing techniques so that you can use them yourself.

DEBATE: Charlie's father has fairly strict rules that he wants his daughters to follow, for example: no makeup, no male company in the house, and no phone calls from boys. What are some of the rules in your house? Daughters: Do you think they are fair? Mothers: What are the reasons for these rules? Choose a rule on which mothers and daughters seem to disagree and hold a debate about it.

REFRESHMENTS OR FOOD MENTIONED IN THE BOOK: During our meeting on this book, our hosts treated us to an authentic Vietnamese dinner. Somehow, food has a way of bringing things to life.

IF YOU LIKED THIS BOOK, TRY...

The Learning Tree, by Gordon Parks—An African-American family struggles to understand and accept the challenges they face in today's world.

Fallen Angels, by Walter Dean Myers—Perry, a Harlem teenager, volunteers for military service when his dream of attending college falls through. This is a coming-of-age tale for young adults set in the trenches of the Vietnam War in the late 1960s.

Some Other Books by Candy Dawson Boyd:

Chevrolet Saturdays
A Different Beat
Forever Friends

Charlotte's Web

by E. B. White

From the moment she sees him, Fern loves Wilbur, a pig born the runt of the litter. She persuades her father to let her raise the little pig, until he is sent to live on a neighboring farm. Here he meets Charlotte, a spider who announces their friendship in her webs, and a host of other barnyard characters. Charlotte and Wibur's unique friendship offers timeless lessons that the humans in their lives also need to learn.

READING TIME: 2–3 hours, about 184 pages
THEMES: friendship, justice, devotion, love, family, imagination, creativity, death

Discussion Questions

- ✦ Why does Fern rescue Wilbur, the runt pig, when her father is about to slaughter him? Why does her father let her keep him? What does her response tell us about her character?
- ✦ What kind of relationship do Fern and Wilbur have? How does Fern show her love for Wilbur, and what does that suggest about the kind of person she is? Have you ever cared for something the way she cares for Wilbur?
- ✦ How does Fern feel when Wilbur has to move to Zuckerman's farm? How does Wilbur feel when he's separated from Fern and lives with the other animals? How is life there different from life with Fern?
- ✦ Why do Charlotte and Wilbur become friends? What are Wilbur's initial impressions of Charlotte? How do his feelings about her change? Would you like to be friends with Charlotte? If Charlotte were a person, what do you think she would be like?
- ✦ How does Charlotte save Wilbur? What do you think about what she does for him? How does it affect Wilbur?

✦ How does Fern feel about Henry Fussy at the beginning of the story? How does that change during the book? What's important about their ride on the Ferris wheel at the county fair?

✦ How do Fern's parents feel about her relationships with the animals in Zuckerman's barnyard? What do their reactions tells us about how they are different from Fern?

✦ What does Charlotte teach Wilbur about friendship? How does Wilbur feel about Charlotte at the end of the story? What's significant about Wilbur taking responsibility for Charlotte's egg sac?

✦ Wilbur considers Charlotte a "true friend and a good writer." What would you want your friends to think about you?

One of the members of our mother-daughter book club, Jamexis, loves stories about animals, so we always have animal stories on our list.

ABOUT THE AUTHOR: Elwyn Brooks (E. B.) White was born on July 11, 1899. In 1921 he graduated from Cornell University, began his writing career as a reporter for a local daily newspaper, and then transferred to an advertising agency. In 1929, he married Katherine S. Angel, with whom he lived in New York. Almost thirty years later, in 1957, he and his wife left New York and moved to a farm in Maine. He is also the coauthor with William Strunk, Jr. of the writing and grammar classic *The Elements of Style.* E. B. White died on his farm in 1985.

Beyond the Book...

ANIMAL STORIES: Pick an animal to research. Learn everything you can about that animal, how they are born, how long it takes them to mature, what they eat, where they live, and so on. Then write a short story about raising that kind of animal.

FARM: Visit a local farm to see how the animals live, or wait until the summer or fall season to attend a fair like the one described in the book.

SPIDERWEB: Design your own spider's web by making one either with thread and yarn or by drawing one. Have it spell out words that are meaningful to you. Or create your own version of Charlotte in her web, with cloth and felt.

WORD POWER: The words in Charlotte's web tell of her feelings for Wilbur. Make a list of words that describe your best friend and your feelings for her.

MOVIE: Rent the animated version of *Charlotte's Web*, narrated by Debbie Reynolds.

REFRESHMENTS OR FOOD MENTIONED IN THE BOOK: Serve a dinner of hamburgers with frozen custard for dessert. Or make a brunch of doughnuts and canned peaches.

IF YOU LIKED THIS BOOK, TRY...

Stuart Little, by E. B. White—This story chronicles the cross-country travels of a determined mouse.

Babe, the Gallant Pig, by Dick King-Smith.

Shiloh, by Phyllis Reynolds.

Some Other Books by E. B. White:

Trumpet of the Swan

Chasing Redbird

by Sharon Creech

In this rich, bittersweet, coming-of-age novel, 13-year-old Zinny Taylor feels like another face in the crowd of her large, loving, but intrusive family. Meanwhile she is dealing with twin pains: her deeply buried grief for her beloved Aunt Jessie, the Redbird of the title, and the long-ago death of her baby cousin. Zinny feels that she needs to do something important, something that belongs to her. She takes on an immense, solitary project—clearing a mysterious trail at the end of which she finds her own true self.

This is a wonderful story of love, loss, and understanding. Sharon Creech is masterful in her ability to build strong characters and deal with serious issues in a charming and often humorous way.

READING TIME: 3–4 hours, about 260 pages
THEMES: grief, separation, death, family, nature, identity, privacy, determination

Discussion Questions

- "Life is a bowl of spaghetti...every now and then you get a meatball." Discuss the spaghetti and meatballs aspects of Zinny Taylor's life. Then, talk about the spaghetti and meatballs aspects of your own life.
- Discuss Zinny's conflicted feelings about her large family. What do you think are the advantages and disadvantages of large families? Of small families?
- Zinny finds a private place to sort out her feelings on the old trail behind the Taylor farm. Have you ever had a private place where you went to sort out your thoughts and feelings?
- Zinny says: "There was something about the trail—I couldn't have said what—that was suddenly so important to me that I became determined to defend it." Why does the trail become increasingly

important to Zinny? Why do you think the author chose for Zinny's challenge an overgrown path rather than a river, ocean, road, or building?

- Zinny is an enthusiastic collector. What do her collections of bottle caps, stones, pencils, key chains, and many other items tell you about her?
- Why does Zinny blame herself for the deaths of Baby Rose and Aunt Jessie?
- Zinny needs quiet and space to deal with the painful memories of her baby cousin, Rose. She says: "One of my teachers once said that we can't get at very early memories because our brains file memories by words, and when we're infants, we don't have enough words." Discuss some of Zinny's memories of Rose, Aunt Jessie, and Uncle Nate. Think back as far as you can. What is your first memory?
- Zinny is certain her mother can't tell her apart from the rest of the family. To this, her mother responds: "If I were blindfolded and you walked in the room, I'd know it was you...Because I know who Zinny is. I know what she sounds like, smells like. I know what she—radiates." Mothers: If you were blindfolded how would you recognize your children? What qualities do your children radiate that make them unique?
- Do you think birth order makes a difference in personality? Does it in this book? If you have siblings, are you the oldest, middle, or youngest, and how do you feel that's affected you?
- What do you think of Jake Boone? How do you feel about his multiple thefts on Zinny's behalf?
- Zinny makes a wish for her Uncle Nate: "I wanted to put Time on a big reel and wind it back so that the three of them [Rose, Aunt Jessie, and Uncle Nate] could be together again." Describe an experience you would like to put on a time reel and play back again.
- In response to Mrs. Taylor's praise for clearing the trail—*"Good for you"*—Zinny thinks: "This had a strange effect on me. Had I actually done something good? Or had something good happened to me?" Answer these two questions.

✦ Zinny's contact with nature and the peace, space, and quiet she finds outdoors help her to sort out her buried feelings of grief. In what way does the natural world help Zinny to find herself?

ABOUT THE AUTHOR: Sharon Creech divides her time between England and the United States, spending nine months in Surrey, England, and the other three enjoying the scenery on Lake Chautauqua in western New York State. She currently teaches literature at an American school in England. One of her greatest pleasures is spending time with her two adult children, Karin and Rob.

Beyond the Book...

COLLECTING: Zinny Taylor is a passionate collector of found objects. Discuss why many girls so enjoy gathering, organizing, and displaying such collections. After the book discussion, invite mothers and daughters who are collectors to show and talk about their passion for collecting and ask them to offer advice to budding collectors on getting started. Mother and daughter pairs might enjoy starting a new collection together from items they have collected on family travels, while pursuing family hobbies, or sorting through family possessions.

FLOWERS: Chasing Redbird is rich with the colors and scents of flowers. Zinnia Taylor beautifies her incredible trail by planting the seeds of her namesake flower, the zinnia. She pays tribute to her baby cousin, Rose, every time she smells a rose. Planting flowers and tending plants connects us to the earth. Plant some flower seeds in a garden or in a pot to celebrate your relationship. Do research on the historical significance and meaning of the flowers you plant. If you want, take plain clay pots to plant your flowers in and paint them with bright colors. Be creative in your designs.

VOLUNTEER: Some outdoors groups, such as the Appalachian Mountain Club, the Sierra Club, or even the Girl Scouts, organize trail maintenance projects. Outdoor lovers might want to investigate opportunities to do volunteer work on local trails. Afterwards, discuss how the experience compares to what Zinny Taylor experienced while she cleared her trail.

BIRD WATCHING: Talk to bird watchers about their hobby. Plan a bird-watching expedition in a nearby park, or even your own backyard. Take a field guide out from the library for guidance. During the outing, talk about what the "redbird" symbolized to the characters in Sharon Creech's book and what birds symbolize to people in general.

REFRESHMENTS OR FOOD MENTIONED IN THE BOOK: A spaghetti and meatballs dinner is a great meal to have after you read *Chasing Redbird.* As everyone helps prepare and enjoy the meal, discuss how life is—or isn't—like spaghetti and meatballs.

IF YOU LIKED THIS BOOK, TRY...

The Remembering Box, by Clifford Eth—A young boy remembers his cherished grandmother when he looks at a special box she bequeathed him.

Walk Two Moons, by Sharon Creech—Another memorable young girl undertakes a journey into her painful past. A Newbery Medal winner (see p. 278).

Some Other Books by Sharon Creech:

Absolutely Normal Chaos

Pleasing the Ghost

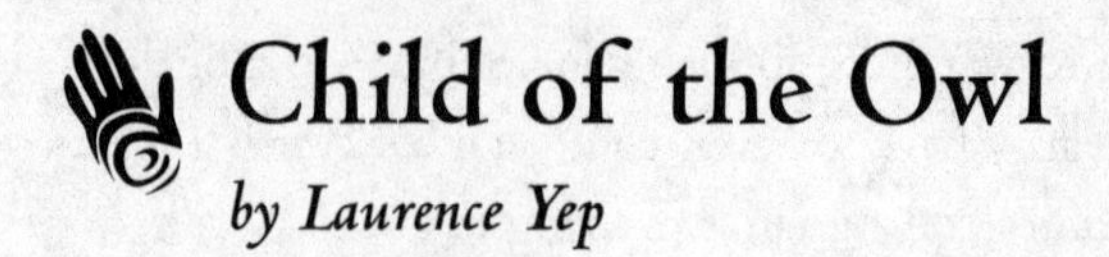

Child of the Owl

by Laurence Yep

A 12-year-old, Chinese-American girl's extended visit with her maternal grandmother forces her to confront fundamental truths about herself and her family. In the process, she arrives at a richer understanding of her identity.

This book broaches the subject of assimilation, a really important topic that is not often made distinct from the theme of racial identity in other novels. I've had the pleasure of meeting Laurence Yep, and I could listen to his stories all day.

READING TIME: 2 hours, about 215 pages
THEMES: assimilation versus ethnic identity, generational conflict, loyalty, responsibility, loss, friendship, father-daughter relationships

Discussion Questions

✦ Barney and Casey don't seem to have a typical father-daughter relationship. Describe how it's different. How does Casey feel about her father's schemes?

✦ How does Barney's stay in the hospital affect Casey? Have you ever had a sick parent, or a parent who had to be in the hospital? How did you feel?

✦ Uncle Phil and Aunt Ethel don't exactly welcome Casey with open arms, and they criticize her father almost constantly. How does that make Casey feel? How would you feel if your relatives spoke that way about your parents?

✦ The issue of responsibility is an underlying theme in this book. What does responsibility mean to Casey's uncle? To her father? What does it mean to you?

✦ Casey observes that "Instead of ordering the twenty-dollar dresser like you'd find in one of our rooms, Phil ordered the two-hundred-dollar dresser. Not because he liked it, but because it had the right price tag." What does Casey think of her uncle's values? How are they different from her father's? From hers?

- ✦ What kind of impression does Casey have about her grandma, Paw-Paw, from her aunt and uncle? How is that different from Casey's response when she finally meets her grandmother?
- ✦ Paw-Paw tells Casey, "The face is the map of the soul." What does she mean by that? Do you think that's true?
- ✦ People make assumptions about Casey that she resents. Have you ever been in a similar situation? How did that make you feel?
- ✦ Casey doesn't identify herself as Chinese to begin with. She says, "I felt like someone had made a mask out of the features of their face and glued it over my real one so everyone would think I was as stupid and mean as those girls outside...those Chinese girls." What changes her feelings about her ethnicity?
- ✦ Why is it important that Casey learn how to eat with chopsticks?
- ✦ When she and her grandmother start spending time together, Casey initially is torn between her assimilated upbringing, and her grandmother's emphasis on Chinese values and heritage. Sometimes people, like Casey's Uncle Phil, feel they have to give up who they are or where they came from to be "American." Do you think that's true? Have you had to give things up to feel like you fit in?
- ✦ Why are the fuschia blossoms in the public garden important to Casey? Have you ever received a gift that mattered to you so much?
- ✦ Casey's father Barney didn't have much to give her except a positive attitude about life. She realizes that he "had a knack for making me see the good side of things." How did he do that? Do you know people who make you feel like that? How are they important to you?
- ✦ How do Casey's experiences in Chinatown, living with Paw-Paw and absorbing her Chinese stories and superstitions, bring her closer to her dead mother, Jeanie?
- ✦ Paw-Paw tells Casey, "You can't be me. You can't be your Mommy, either. You can only be yourself." What is she trying to tell Casey?
- ✦ What kind of relationship does Casey have with Paw-Paw? How

does Paw-Paw help Casey find her identity? Have you had that kind of relationship with a grandparent or other relative?

- ✦ Casey observes, "Just because a thing is old doesn't make it lose its meaning." How does this relate to Paw-Paw? To Casey's Chinese heritage?
- ✦ Throughout the novel, Casey compares herself to the owl, referring to the stories that her grandmother has told her. What does the owl represent to Casey? What significance do these stories hold for her?

ABOUT THE AUTHOR: Laurence Yep was born and raised in San Francisco, CA. In conversations about his youth, as one of only a few Asian-Americans in his area, Yep has said "I was the all-purpose Asian. When we played war, I was the Japanese who got killed...when the Korean war came along, I was a North Korean Communist." Yep gathers these experiences and others as a basis for many of his books. Yep was hooked on writing after an assignment that required he have work published in a national magazine. He did not actually have anything published until he was in college, and was once paid a penny a word for a story he submitted to a science fiction magazine. Yep currently lives in Sunnyvale, CA.

Beyond the Book...

CHINESE-AMERICAN EXPERIENCE: Make a presentation on the Chinese-American experience in America, including the immigrant experience and the development of distinctive ethnic communities. Talk about what brought the Chinese to America, and ask members of the group to discuss why their forebears immigrated, and how their experiences may have been similar or different from that of Paw-Paw.

MEMORY BOOK: Ask your grandparents or an older neighbor to share stories about their ethnic heritage that you can keep in a special memory book.

CHINATOWN: Visit a nearby Chinatown community, if there is one. Eat out at a Chinese restaurant, and use chopsticks.

CHARM: Try to make your own significant charm, like Casey's owl charm, and tell the group why you chose your particular symbol.

REFRESHMENTS OR FOOD MENTIONED IN THE BOOK: To suggest the atmosphere of the book, serve a variety of Chinese dishes for refreshment, especially Chinese tea and the special buns that are considered a treat.

IF YOU LIKED THIS BOOK, TRY...

The Secret of Roan Inish, by Mick Lally—This book is about a special relationship between grandparents and grandchildren. It also deals with the conflict between the modern, assimilated world, and the older, more superstitious one represented by the grandparents.

Older readers might want to try *The Joy Luck Club,* by Amy Tan—This book also deals with issues of Chinese-American identity.

Some Other Books by Laurence Yep:

Butterfly Boy

City of Dragons

The Children of Green Knowe

by L. M. Boston

In this haunting fantasy, Tolly, a lonely young boy, is sent to live with his great-grandmother at the family's ancestral home. As she tells him stories of his family's history, he is introduced to a magical world that provides him with unusual friends, and the feeling of belonging he has been yearning for.

I was very fortunate growing up to have had wonderful relationships with my grandparents. I loved my grandfather, Wawa, dearly and still mourn his loss. This book reminded me how magical and special grandparent-grandchild relationships can be.

READING TIME: 1–2 hours, about 182 pages
THEMES: loneliness, imagination, family, friendship, loss, grief, aging

Discussion Questions

- What is Tolly's family like? How does he feel about his father and stepmother? What kind of family does he yearn for? Why?
- Tolly has many fears, but is reluctant to reveal them. Why? Have you ever been in a situation where you felt the same way?
- What are Tolly's first impressions of Old Knowe? Why does Tolly feel as if he knows her "quite well" even though they just met?
- What does "home" mean to Tolly? What does it mean to you? Why does Green Noah become important to Tolly?
- What do Tolly and his grandmother share that creates a strong bond between them? If you have had that kind of relationship with a grandparent, describe what makes the experience special.
- "Both their faces were bright with excitement that for all the difference in their ages they were very much alike." What made Tolly and his grandmother realize they were so similar? How are they alike?

- Why are the ghost children, Toby, Alexander, and Linnet, important to Tolly?
- How does Tolly feel about the curse on Green Noah?
- Why is the gift of Alexander's flute so significant to Tolly?
- Even though the children are vivid playmates for Tolly, he has to confront the reality that they died centuries before. How does that change how Tolly feels about death? What are your feelings about death? Has anything happened in your life that's influenced them?
- Tolly ultimately recognizes that the ghost children "were like brothers and sisters who come and go, but there is no need for worry; they are sure to come home again." What does Tolly mean by that?
- How does his stay in Green Noah and meeting his grandmother and all of the ghost children change Tolly?

ABOUT THE AUTHOR: Lucy Maria (L. M.) Boston was born in England in 1892. In 1935 she purchased a run-down manor house near Cambridge. She spent two years restoring it, and in doing so, was inspired to write the Green Knowe series. Ms. Boston died in 1990 at the age of 98.

Beyond the Book...

TOPIARY: Topiaries, or specially pruned shrubbery, play an important role in this novel. Go to the library and take out some gardening books that explain how topiaries are created. Try to create a simple topiary animal from a small plant or shrub that can be transported easily.

ENGLISH CIVIL WAR: Prepare a presentation on England's Civil War, with an explanation of the differences between the Royalists and the Roundheads, to understand better the historical context for Alexander and Linnet's experiences.

NOAH'S ARK: The Noah's Ark game is especially meaningful to Toby. Find a set of animals and a toy boat to play with, and talk about what it would have been like to have had only simple toys that rely on a your imagination for recreation.

MOVIE: Rent the video *Fairy Tale,* a fantasy that explores what happens to imaginative children who have a special relationship with the unseen world in a quiet English village.

REFRESHMENTS OR FOOD MENTIONED IN THE BOOK: For refreshments, prepare a traditional English tea, like the kind Tolly enjoys with his great-grandmother. Include buttered toast, sliced egg sandwiches, cold chicken sandwiches, frosted chocolate cookies, and hot tea with sugar or honey.

IF YOU LIKED THIS BOOK, TRY...

The Secret Garden, by Frances Hodgson Burnett—This classic story is about an orphan girl's self-discovery in a mysterious English house.

The Chronicles of Narnia, a series by C. S. Lewis—English children explore a magical land.

Mary Poppins, by P. L. Travers—A fantasy well-rooted in reality.

Some Other Books by L. M. Boston:

The River at Green Knowe
An Enemy at Green Knowe
A Stranger at Green Knowe

Cousins

by Virginia Hamilton

Family life has become complicated for Cammy. Her wonderful grandmother is wasting away in a nursing home. She and her brother seem to be the poor relations of their clan. But for Cammy, the worst thing of all is being cousins with the perfect and accomplished Patty Ann who can seemingly do no wrong. Cammy's secret—and not-so-secret—rivalry with Patty Ann turns into guilt and grief when Patty Ann is killed in a freak accident.

This book started the most provocative discussion our group has had. It was the first time there was any real dissension between the moms, some of whom felt that Patty Ann's snobbish attitude was a cover-up for deep-rooted insecurities and sensitivity, while others felt that she was truly not a nice person. This disagreement was interesting for the girls, who were used to seeing all the moms side with each other.

READING TIME: 2 hours, about 125 pages
THEMES: family, jealousy, body image, death, coming of age

Discussion Questions

✦ Cammy secretly compares herself to Patty Ann and feels like everyone else at home and school does too. What do you find when you compare Cammy and Patty Ann with each other? How are they alike? How are they different?

✦ "To Cammy, everything [Patty Ann] did was like chalk scraping on a blackboard." What are some of the things Patty Ann does that irritate Cammy?

✦ The story mainly focuses on Cammy's own feelings about herself. How do you think Patty Ann feels about Cammy?

✦ What are Cammy's feelings about her grandmother, Tut?

- ✦ Describe Cammy's relationship with her brother, Andrew. How are Cammy and Andrew alike?
- ✦ To all appearances Patty Ann is accomplished, secure, poised, self-disciplined, and intelligent. What hints are there that she has her share of problems?
- ✦ What makes Cammy likable enough for the reader to accept that she can be petty and jealous?
- ✦ Have you ever been jealous of someone the way Cammy is jealous of Patty Ann? What does jealousy feel like? What are some positive ways people can get over feelings of jealousy?
- ✦ Describe Cammy's relationship with her father. How does her mother relate to Cammy's father? Why does he come back?
- ✦ What is significant about the episode when Cammy, Elodie, and Patty Ann are making corn husk dolls? Why is it important to Cammy that Elodie left her table?
- ✦ Compare how Cammy felt about Elodie with the way Elodie felt about Cammy. Do you think Cammy should have been more understanding of Elodie's feelings?
- ✦ When she is talking to her dad upon his arrival home, Cammy admits to having seen Patty Ann's face while she was drowning. Why didn't she want to admit that she had seen it?
- ✦ What makes Cammy unable to get up and go to school after Patty Ann's death? Why does Patty Ann appear in her dreams? What finally helps Cammy get back into the stream of life again?
- ✦ The last line of the book is: "I get it, now." What is it that Cammy now gets?

ABOUT THE AUTHOR: Virginia Hamilton was born and raised in Yellow Springs, OH, where she still lives with her husband, Arnold Adoff, an award-winning poet. Her hobbies include swimming, walking, surfing the Internet, and reading. She describes *Cousins* as semi-autobiographical—she, too, came from a large family with lots of cousins, lived on a farm in the country, and went to day camp like the girls in this book do. She also nearly drowned once and thinks that some of the horror of that situation might have contributed to her vision of Elodie and Patty Ann's encounter with the Bluety. Hamilton is an accomplished writer who draws from her African-American and Native American roots in her books. She and her husband have two children, Leigh and Jaime. Hamilton has been awarded every major award and honor given to American authors of literature for youth.

Beyond the Book...

FAMILY TREE: Before the book discussion of *Cousins,* create a family tree or look at a family photo album. Share memories of family gatherings, rivalries, and reconciliations. Think about what identifies your family as a family beyond blood ties. Does your family have a sense of mission or particular values and ways of doing things that are unique?

CORN HUSK DOLLS: Make corn husk dolls like the ones Cammy, Elodie, and Patty Ann made at camp. Reread the passage from the book where Ms. Devine is telling the class how it should be done and compare how yours comes out to the way Patty Ann's seemed to have come out. How would you have felt if you had been in Cammy's position?

FAMILY TRADITIONS: Cammy's relatives are a huge part of her life. To strengthen family bonds, invite relatives, like grandmothers, aunts, and cousins, to talk about *Cousins* with you. Encourage these guests to join in a discussion of their family's rituals, values, lore, and tradition.

VOLUNTEER: Visit a nursing home. Brainstorm how you might help elderly people during your visit, perhaps by reading to them, organizing a mini-concert, serving a meal, or playing games. Or call a local nursing home and ask whether it would be possible to host an ice cream and cake party (Cammy's favorites) for the residents.

REFRESHMENTS OR FOOD MENTIONED IN THE BOOK: Ice cream and sheet cake are two of Cammy's favorite family foods. Serve ice cream and cake as part of the book discussion of *Cousins.*

IF YOU LIKED THIS BOOK, TRY…

Sister, by Eloise Greenfield—This coming-of-age novel deals in part with sibling rivalry and feelings of guilt.

Nobody's Family Is Going to Change, by Louise Fitzhugh—This humorous novel of a middle-class black family tells of the daily ups and downs of its members.

Some Other Books by Virginia Hamilton:

The House of Dies Drear (see p. 136)

Her Stories: African American Folktales, Fairy Tales, and True Tales

M. C. Higgins the Great

The Cuckoo's Child

by Suzanne Freeman

For 11-year-old Mia, the return to America from Lebanon is both a dream and a nightmare. Homesick for the United States after living several years in Beirut, Mia comes home with her older stepsisters but without her footloose, loving, nonconformist parents who are presumed lost at sea in Greece. Mia searches the horizon every day for signs of their return while also searching deep inside herself to discover who she really is without her parents.

READING TIME: 3 hours, about 250 pages
THEMES: loss, guilt, family, homesickness, coming of age

Discussion Questions

- Why is this book called *The Cuckoo's Child*? Why does Mia dislike the cuckoo bird?
- Why is Mia drawn to high places, both in Beirut and in Tennessee?
- Mia's homesickness for America is very real and specific. She misses saying the Pledge of Allegiance, eating corn on the cob, and shopping at variety stores. Discuss any experiences of homesickness that you've had. What specific things from home did you miss? Did this longing feel like an actual sickness? How did you get over it?
- When she is in Beirut, Mia longs for the sameness of America. "There was one big America, and what I loved about it the most was that it was all the same." Do you agree or disagree with Mia's statement?
- Mia has an up-and-down relationship with her two stepsisters. What do you think is difficult about stepsibling relationships?
- What are Mia's mother's views about religion? How are her views different from Mia's feelings about religion?

- What does Mia remember about her mother? What do you think of Mia's mother as she remembers her?
- Mia recalls periods of wanting her mother to do things for her that she could do herself: "...cut the bruises out of an apple, run my bathwater, look through the laundry basket for my Girl Scout kerchief." Discuss whether you have gone through similar periods of letting yourself be babied by your parents. What specific tasks have you had them do instead of doing them yourself? What is behind this impulse of being dependent about certain things?
- In contrast to her mother, Mia believes: "You followed the rules, and you got to live a regular life. I would never understand why people wanted anything else." What rules is Mia talking about? What is a "regular life" in her opinion? What "else" might people want? Are your life expectations more like Mia's or her mother's?
- Mia recalls telling her Beirut schoolmates all about America, but she doesn't share her Beirut experiences with her American classmates. Why?
- Were you surprised when Mia turned mean and dishonest? How did you feel about this change in her? What do you think brought on this change?
- Did you expect the book to end with Mia's parents being found? How did you feel about the ending?
- How does Mia feel about her own individuality by the end of the book?

ABOUT THE AUTHOR: Suzanne Freeman spent some of her childhood in Beirut, which influenced her writing later in life. During this time she found that she missed the American things that she had become accustomed to as well as the life that she had in the United States. These feelings and memories play a large role in *The Cuckoo's Child,* since it is about a young girl who also lives abroad. Freeman has taught writing at the University of Virginia, worked at a newspaper, and currently reviews books. She and her husband have two children and live in Manchester, MA.

Beyond the Book...

LISTS: In Beirut Mia misses very specific things about America—everything from Saturday movie matinees to drinking milk from a carton. Make lists of the top ten things you would miss about the United States. If you were not born in the United States, list what you would miss most about your home country.

POSSESSIONS: A few years ago, some friends of ours moved to France for a year so their three children could learn French. In the process of getting ready to leave, they were forced to sort through all of their material possessions to decide what was important enough to take with them. The conclusion they came to was that they had "too much stuff," so they chose what was important and what they could do without. I think this would be an interesting exercise for us all.

Suppose you were going to live in Beirut, or France, or some other faraway place for a year and you could only take a few items. Make a list of what you would take and why. Talk about your choices.

REFRESHMENTS OR FOOD MENTIONED IN THE BOOK: American favorites like sloppy Joes, hot dogs, and corn on the cob are mentioned longingly in this book. Prepare and share a typical all-American dish—or a typical ethnic dish if you are not American.

IF YOU LIKED THIS BOOK, TRY...

Homesick: My Own Story, by Jean Fritz—An expatriate American girl living in China longs for an America she has never seen (see p. 132).

Walk Two Moons, by Sharon Creech—A young girl sets out on a journey to find her missing mother and in the process discovers who she is herself (see p. 278).

The Dark Side of Nowhere

by Neal Shusterman

Jason is bored living in a small town—until the sinister school janitor, Mr. Grant, gives him a weird glove that can shoot steel pellets with deadly force. Soon, Jason learns that other kids have gloves like his own, and that all of their parents came to earth from another planet. In fact, the monthly "allergy" shots they all get are what keeps them looking human. Now their shots have stopped, and Grant is training the children to take over the Earth and become the "master race." Jason, human at heart if no longer in body, foils the plan in the end.

READING TIME: 3–4 hours, about 185 pages
THEMES: identity, leadership, friendship, family, race

Discussion Questions

- Why does Jason like Paula so much? Why is Jason so moved when Paula gives him her baseball cap? Would some boys feel differently about Paula's gesture? Who would you rather have as a boyfriend—someone like Jason, or a boy with more old-fashioned ideas about boys and girls' roles in romantic relationships? Talk about your reasons.
- Jason says that he has started thinking of himself as "one of us instead of one of them." Soon he comes to believe Grant's words: *"We are magnificent. . . . We are above and beyond all others."* Why is it that, "Deep down everyone wants to believe something like that. . .?" What are the dangers of this type of thinking?
- Jason tells the group, "We turn whole social groups against one another. Keep them hating; keep them divided." Later, Ethan quotes Grant's words: ". . .we shouldn't think of them (humans) as real people, like us." Talk about some of today's world problems that result from one group turning against another; from one group hating another; from one group thinking of another as subhuman.

- ✦ Grant encourages the group to feel "superior" to humans. At one point, he asks them, "Would we have been created so beautiful if we weren't meant to have the world?" Do you think the author wants the story to remind you of people and events in history or in the news? If so, what events do you think he's referring to? Do you think this book has a message?

I couldn't put this book down. It has a great story, but more important, it deals with interesting, timely issues, and really expands a reader's mind.

- ✦ Jason says that when you live only with people like yourself, "you've got nothing to feed on but the same thoughts and ideas bouncing back at you from your friends...things that start to sound normal and reasonable have no bearing on what's true." How can Jason's words be used as an argument for living in a diverse community, or going to a school with people from different groups? Why do you think more people don't choose to live in diverse communities?
- ✦ Jason's friend Wesley betrays him to Grant, even though he doesn't want to. Why does Wesley turn Jason in? What does Wesley mean when he tells Jason, "I'll never be like you. I'm not that strong"? What kind of strength does Wesley lack? Why is it important to develop that kind of strength, and what are the best ways to do it?
- ✦ Jason's last words in the book are, "I am human." Jason's parents are from another planet, and Jason does not look like a human being. Why does he say he is human? What does being human really mean?
- ✦ Do you believe there might be life on other planets or elsewhere in the universe? Why or why not?

ABOUT THE AUTHOR: When Neal Shusterman was growing up in Brooklyn, NY, he began writing as an alternative to playing stickball. Books always played a large role in Shusterman's life. As a child he was even able to figure out the publisher of a book by the smell of the paper and the ink. His professional career began soon after he graduated from college. Shusterman has been successful as a columnist, novelist, screenwriter, and television writer.

Beyond the Book...

PORTRAITS: Draw pictures of what you think Ethan, and later, Jason, looked like. Use Shusterman's descriptions as a guide. You might want to draw portraits of other characters as well—Billy, for example, or Paula. The author provides ample descriptions of their physical characteristics.

MOVIE: After Jason's father tells him where his family came from and who they are, Jason remarks, "So...we're body snatchers?" Watch a videotape of the film *Invasion of the Body Snatchers.* Make a comparison-and-contrast chart showing similarities and differences between the film and *The Dark Side of Nowhere.* Write a "What if?" paragraph on both the movie and the book, speculating what would happen *if* each plot took a different turn and the extraterrestrials were successful in wiping out or subjugating the human race.

IF YOU LIKED THIS BOOK, TRY...

The Time Quartet, by Madeleine L'Engle—In *A Wrinkle in Time* (see p. 314), the first book in the quartet, a brother and sister embark on a perilous quest through space to find their father and battle a cosmic evil force. The other three titles are: *A Wind in the Door, A Swiftly Tilting Planet,* and *Many Waters.*

"Dandelion Wine"(short story), by Ray Bradbury—A 12-year-old boy discovers that his small town holds strange stories.

Childhood's End, by Arthur C. Clarke—A novel of the next stage in the evolution of humanity.

Some Other Books by Neal Shusterman:

S.P.E.E.D.I.N.G. B.U.L.L.E.T.

The Eyes of King Midas

The Dark Stairs: A Herculeah Jones Mystery

by Betsy Byars

Nosy, independent, and determined, Herculeah Jones just can't help getting involved in the family business—solving mysteries. Though her parents are divorced, Herculeah is close to her detective father and private investigator mother. When some strange events transpire at a local abandoned house, Herculeah is determined to figure out what is going on, no matter what the danger.

READING TIME: 2 hours, about 130 pages
THEMES: mystery, mental illness, divorce

Discussion Questions

- ✦ What does Hurculeah get out of her partnership with Meat? Why do you think so many strong, fictional, detective heros often have a bumbling partner just as Herculeah does with Meat? Can you think of examples of this kind of partnership in some well-known mysteries?
- ✦ What words would you use to describe Herculeah Jones? What qualities does she have that would make her a good detective?
- ✦ Have you ever gone through a phase where you wanted to become a spy or detective? Why do you think these jobs appeal to certain kids?
- ✦ What's significant about the names of people and places in this book? What do names like Herculeah, Meat, Mr. Crewell, and Deak Oaks connote?
- ✦ There's a fair amount of spying, tape recording, and eavesdropping in this book. Do you think these are the only ways to find out what people are really all about? How else can you find out about people?

- ✦ Describe Herculeah's relationship with her parents. Why doesn't Herculeah's father want her to become a detective?
- ✦ Have there ever been any mysterious goings-on in your neighborhood? Discuss them with the group. Is there any house in your neighborhood or town that people consider to be haunted? Why?
- ✦ When did you figure out what this mystery was all about? What clues gave you an idea about how it would turn out?

ABOUT THE AUTHOR: Betsy Byars was born in Charlotte, NC, in 1928 and began her writing career when she was a mother of four young children with the submission of a short piece to the *Saturday Evening Post.* Her first books weren't written, though, until her children were in their teens. These days, she does most of her writing in a log cabin not far from her house. Byars has a pilot's license and when she isn't working, she shares her husband's passion for glider flying.

Beyond the Book...

MYSTERY: Ask mystery lovers in the book group to make up a list of their favorite mysteries and movies for other interested readers in the group. Invite mystery lovers to discuss what it is about mysteries that they like and how they first got interested in the genre. Calling all mother and daughter Nancy Drew fans: Compare Herculeah Jones and Nancy Drew—both the heroines as well as the books. Are there differences between what moms and daughters love about mysteries?

DETECTIVE: Track mysterious goings-on in your neighborhood, like missing trash cans or the hijinks of a neighbor's cat, and start a journal that records every strange occurrence. Write a story based on those happenings and what you imagine might be going on.

BOOK SWAP: Have a mystery book swap—girls with girls and moms with moms—after the book discussion so that mystery lovers can indulge their habit with books they haven't read.

IF YOU LIKED THIS BOOK, TRY...

Harriet the Spy, by Louise Fitzhugh—Another independent-minded girl explores mysterious goings-on in her neighborhood (see p. 122).

Anastasia Krupnik, by Lois Lowry—This is the first in a series of books about an adventurous girl solving local mysteries.

Some Other Books by Betsy Byars:

After the Goatman
Cybil War
Good-bye, Chicken Little

A Day No Pigs Would Die

by Robert Newton Peck

"Pinky" is Rob's beloved pet pig, given to him as a reward for delivering twin calves and saving the mother cow's life. As Pinky becomes full-grown, however, circumstances dictate against keeping her. This autobiographical novel tells the story of a Shaker boy growing up on a Vermont farm, and the joys and heartaches that mark his passage into manhood.

I chose to include this book not only because it is a classic and important work, but because of its portrayal of a father-son relationship. The male point of view makes it easier to address issues that might otherwise be awkward for mothers and daughters to raise together. This book also gives girls some insight into how boys think.

READING TIME: 2–3 hours, about 150 pages
THEMES: coming of age, family, difference, grief, loss

Discussion Questions

- As the book begins, Rob is cutting school because another boy made fun of his Shaker clothes. What defines Shaker clothes? How are they different? Have you ever had an experience in which one person made fun of another's clothing or appearance? How does the person who is made fun of feel? What makes some people want to make fun of others?
- Letty Phelps, Haven Peck's cousin, was an unwed mother. After both she and her baby are dead, the Pecks' neighbor, Sebring Hillman, finally admits to being the baby's father. Why do you think the author chose to include this episode in the book?
- Based on how Rob's father acts when Rob has been hurt and when he returns from the Rutland Fair, what do you think of Haven Peck as a father? Do you think he loves Rob? Why is he so stern? What values does he try to pass down to Rob?

- ✦ Do you know your neighbors? What do you think makes a good neighbor? Why is it important to be a good neighbor in the Shaker community?
- ✦ How is Rob's mother different from her husband? What do you think of her as a mother? What values will Rob learn from her? Find examples in the book to back up your opinions.
- ✦ Rob's father tells him that by the time he is 13, he will have to be a man, the head of the family. Have you ever taken on a responsibility you didn't necessarily feel ready for? How did it make you feel? Did it have a positive or negative effect on your life?
- ✦ Rob is surprised to discover that Mr. and Mrs. Tanner and Aunt Matty are Baptists, not Shakers, like himself and his parents. Why does this surprise him? What makes it possible for people with different religious, cultural, or political ideas to get along?
- ✦ How did you react to Rob's father when Pinky had to be butchered? Do you agree with the way Rob's father handled the situation? For example, should he have done the butchering when Rob wasn't home? Should he have allowed Rob to keep Pinky? Give reasons to back up your opinions.
- ✦ When Rob's father dies, Rob says, "We all would long for a different parcel of him." What "parcel" of his father will Rob long for? What is most important to you about each of your parents?

ABOUT THE AUTHOR: Robert Newton Peck has written over 60 novels; 13 have been bestsellers. Peck grew up in rural Vermont, which provided inspiration for many of his novels. Peck now lives in Longwood, FL.

Beyond the Book...

SHAKER RESEARCH: Find out more about the Shakers. When was the sect founded? What are their beliefs and practices, in addition to those mentioned in the book? What are their contributions to American life? (The circular saw, cut nails, washing machine, flat brooms, metal pen points, and finely designed furniture are a few!) You might request information from the Shaker village in Harrodsburg, KY, or the

Sabbathday Lake Community in Maine. Also, if you have access to the Internet, there is a wealth of information available, using the search word "Shakers."

MAP: Especially for those not living in the northeastern United States, find out more about the part of the country where the Pecks lived. Locate Rutland, VT, on a map. In a book about the New England states, or from brochures obtained from the Vermont Chamber of Commerce, find pictures of the rural landscape around Rutland. Encourage the girls to draw or paint pictures of a similar landscape.

HYMN: Listen to a recording or learn the Shaker hymn "Amazing Grace." Practice singing it together.

REFRESHMENTS OR FOOD MENTIONED IN THE BOOK: When Rob comes home from the Rutland Fair, his mother has a homemade blackberry pie waiting for him. Make your own homemade pie. If you can't find blackberries, use fresh blueberries or another kind of fruit. Make and roll the dough yourselves. Enjoy the different stages of making your pie, the wonderful smell coming from the oven as it bakes, and of course, the delicious taste when you eat it.

IF YOU LIKED THIS BOOK, TRY . . .

A Part of the Sky, by Robert Newton Peck—This is a sequel to *A Day No Pigs Would Die.*

The Yearling, by Marjorie K. Rawlings—A growing-up story of a young boy who, like Rob, must lose his beloved pet—a deer.

Some Other Books by Robert Newton Peck:

Arly

A Part of the Sky

Soup

Dogsong

by Gary Paulsen

This coming-of-age adventure novel tells the story of a 14-year-old Eskimo boy who is alienated from his father and from the life of modern Eskimos. The boy longs for meaningful challenges that contemporary Eskimo life fails to provide. Only when he learns the lessons of the "old ways" of Eskimo life from an Eskimo elder does the boy finally become a real man in his ancient culture.

READING TIME: 2–3 hours, about 178 pages
THEMES: cultural identity, death, survival, manhood, self-discovery

Discussion Questions

✦ Describe the relationship between Russel Susskit and his father early in the book. Why does Mr. Susskit allow Russel to live with Oogruk?

✦ Why can't Mr. Susskit pass on Eskimo wisdom to Russel? How does that affect their relationship?

✦ In a reversal of the usual situation of a son or daughter rebelling from older family customs, Russel embraces the old way of Eskimo life that his father has cast off. Describe some of your family's old-fashioned customs and each family member's attitude about these customs. Is there anyone you know who has become more traditional than their parents?

✦ Russel's father senses his son's confusion about growing up. "It is part that you are fourteen and have thirteen winters and there are things that happen then which are hard to understand." What "things" do you think he's talking about? What kinds of confused feelings do you think 14-year-olds have?

✦ Oogruk tells Russel that the old Eskimo songs are disappearing. Why are these ancient songs important? What purpose do they serve in the Eskimo culture?

- ✦ How does Oogruk feel about the animals a true hunter must kill in order to survive? How do the traditional Eskimo hunters show respect for the animals they kill?
- ✦ How does Russel establish a relationship with Oogruk's sled dogs? What does he do to assert his authority over these magnificent working animals?
- ✦ "Russel…was not rebelling. He was working toward something in his mind, not away from something he didn't like." What is the distinction being made in that statement? Do you think rebellion is an action *against* or *towards* something? Can you provide examples from your own life to illustrate your views?
- ✦ Why did Oogruk want to die the way he did? What do you think of Oogruk's chosen way of death?
- ✦ The author repeatedly emphasizes the word *out* and often compares the Eskimos' indoor life to their outdoor life. How are the two environments different in *Dogsong*?
- ✦ Russel experiences several dreamlike visions. What messages do his dreams send him? Do you believe that dreams contain messages?
- ✦ Even after Oogruk's death, Russel thinks about the old man's words of wisdom: "It isn't the destination that counts. It is the journey. That is what life is.…Pay attention to the journey." Discuss what this quotation means. How does it affect Russel?

ABOUT THE AUTHOR: Gary Paulsen has a somewhat different approach to writing novels—he has to live what he writes. His novel *Dogsong* was inspired by his participation in the Iditarod (an Arctic dogsled race that spans 700 miles). He currently lives in Becida, MN.

Beyond the Book…

ALASKA: Prior to the book discussion on *Dogsong,* take out library picture books about Alaska, particularly those that feature pictures of its native peoples, sled dog teams, and the tundra. Invite members who may have visited or lived Alaska (or a guest who has) to talk about this faraway part of the country.

DOG TRAINING: Through training their own dogs, dog owners will have some idea why Russel must establish his dominance over Oogruk's dog team—that this establishment of "top dog" authority is part of every dog's wolf-pack ancestry and necessary for the pack's survival. Ask dog owners to explain how they have trained their dogs to obey them.

VOLUNTEER: Spend some time working at a local animal shelter.

MOVIE: Rent and view a video about survival experiences, such as *My Side of the Mountain.* Talk about the ways people handle crisis situations and how you would go about making decisions in that kind of situation. Or rent either the 1933, 1972, 1976, or 1993 movie version of Jack London's *Call of the Wild* to gain an understanding of the bond between a young man and a sled dog.

REFRESHMENTS OR FOOD MENTIONED IN THE BOOK: The Eskimos eat a lot of meat they have hunted themselves. Try making some venison, or go fishing and cook the fish yourself. The Eskimos also eat a lot of raw fish. Serve some sushi to see what it is like.

IF YOU LIKED THIS BOOK, TRY…

Eskimos: The Inuit of the Arctic, by J. H. Greg Smith—This introduces readers to the culture and history of one of the major native Eskimo tribes of the Arctic.

Julie of the Wolves, by Jean Craighead George—Like *Dogsong,* this novel is also the story of a young Eskimo coming of age in the wild—this time from a girl's experience (see p. 157).

Racing Sled Dogs: An Original North American Sport, by Michael Cooper—This account describes the annual Iditarod Sled Dog race in Alaska.

Some Other Books by Gary Paulsen:

Brian's Winter

Canyons

Dragon of the Lost Sea

by Laurence Yep

In this fantasy, two homeless orphans, Thorn, a servant boy, and Shimmer, a dragon princess, combine forces on a quest to reclaim the dragon's lost home, a sea, now contained in a magic pebble. Along the way, Thorn realizes that in spite of their differences he and the dragon have something in common: They are alone in the world and they need each other.

READING TIME: 1–2 hours, about 211 pages
THEMES: loneliness, friendship, justice and injustice, family, prejudice, self-esteem

Discussion Questions

✦ Shimmer, the dragon princess, tells us, "When trouble isn't drawn to me, I seem to be drawn to it." What does she mean? What does that suggest about her character? Do you know anyone like that, and how does that affect you?

✦ How does the children's mistreatment of Thorn affect Shimmer?

✦ Thorn didn't react when the children teased him, but protects and defends Shimmer when the children mock her in her disguise as an old beggar woman. What does that tell you about his character? How do you think you would have responded in a situation like that?

✦ Thorn hadn't been shown much kindness or concern in his life. Where do you think he learned to act generously as he does toward Shimmer?

✦ What kind of relationship do Thorn and Shimmer have? How does that relationship change over the course of the novel?

✦ Shimmer can be vain, boastful, abrupt, even rude and ungracious. Why do you think Thorn is drawn to her?

✦ Why does a strong bond develop between Thorn and Shimmer?

- ✦ As a dragon princess, Shimmer takes the idea of honor very seriously. Is there something about which you feel as strongly?
- ✦ During their fight with the keeper's pets, Shimmer considers Thorn a "brave fool." How does that change their relationship?
- ✦ How do Shimmer's feelings about her brother's treatment of her affect her life? How would you feel if you were treated that way by a sibling?
- ✦ Thorn and Shimmer have very different ideas about families. Why do you think this is? Whose ideas do you identify with? Why?
- ✦ Why doesn't Shimmer abandon Thorn in the salt remains of the Lost Sea? Do you think she would have left him at the beginning of the book? How have her feelings for him changed?
- ✦ Why does Shimmer sacrifice her opportunity for Thorn? What do you think you would have done?
- ✦ What kind of person does Civet, the sorceress who stole Shimmer's home, turn out to be? How do Shimmer's feelings about her change when she learns this? After you hear Civet's story, what do you think about her?

ABOUT THE AUTHOR: (see p. 46)

Beyond the Book...

CHINESE FOLKLORE: Go to the library and do some research on Chinese folklore, mythology, and legends. The dragon is an important figure in traditional Chinese stories, but there are many others as significant in Chinese literature and art. Including some Chinese paintings in the presentation will help you understand and see how this novel extends the tradition.

AMULETS: Amulets, or talismans, are significant in this story. Either make individual blue pebble necklaces (using smooth stones that can be painted blue) like the one Shimmer treasures, or make other meaningful charms. You can take stones or other objects to a jewelry, bead, or craft store to have holes drilled in them, or wrap them in bead wire to make them into pendants.

WATER WALKS: Water is practically a character of its own in this story. As a group or alone, take a walk along a nearby lake, seashore, or river. Make up your own legend about that particular body of water, and share those stories with the group.

REFRESHMENTS OR FOOD MENTIONED IN THE BOOK: Serve Chinese noodles, and assorted fresh fruits like those mentioned in the book—including pomegranates (when in season), plums, pears, dried fruits, and almonds, that are particular treats in the book. Talk about how these fruits, most of which are commonly found in local supermarkets, were considered such delicacies and treats for Thorn.

IF YOU LIKED THIS BOOK, TRY...
The Sword in the Stone, by T. H. White
Some Other Books by Laurence Yep:
Dragon Steel
Dragon Cauldron
Dragon War

The Egypt Game

by Zilpha Keatley Snyder

What starts off as an innocent and imaginative game takes on a more sinister note as a group of independent 11-year-olds find themselves deeply involved in an elaborate fantasy, one that they themselves created.

When I was growing up, I had a special, although not secret, place to play. My grandfather, whom we called Wawa, built a playhouse in the backyard where my friends and I used to play. Looking back, I think we longed for some of the mystery and adventure of *The Egypt Game.*

READING TIME: 2–3 hours, about 215 pages
THEMES: creativity, friendship, family, loneliness, secrecy, popularity, self-esteem

Discussion Questions

✦ How do Melanie and April feel about each other when they first meet? How does that change during the course of the book?

✦ Why is April "showboating" when she first meets Melanie? How does Melanie know she's showing off? Have you ever acted like that? Why?

✦ April is worried about starting school. Have you ever felt nervous about starting a new school year? Why is it especially anxiety-producing for April?

✦ What kind of relationship does April have with her grandmother? How does it evolve?

✦ April is bright, creative, and imaginative, but sometimes she crosses the fine line that separates imagination from lies. What makes her do that? Have you ever been prompted to do something like that? If so, what motivated you and what were the consequences?

✦ Because of her work as an actress, April's mother sends April to live with her paternal grandmother. April cherishes the hope that her mother will send for her, even though it's unlikely. She says, "...a bright and beautiful blur, no matter how distant, was better than a reality that was dull and gray." Do you agree or disagree with this philosophy? Why or why not?

- How does the Egypt game change when Elizabeth, the new neighbor, joins Melanie and April?
- The Egypt game is a special secret the children share. Why is Egypt important to the children? Have you ever had a special place that was as important to you as the Egypt game is to Melanie, April, and the others?
- After a while, strange things begin happening in Egypt, and the children become scared of what they've created. Have you ever started something that ultimately got away from you, and overwhelmed you? If so, what happened and how did you resolve the situation?
- April defines friendship like this: "You didn't pick your friends just because they were handy—or even lonely. You picked them because you thought alike and were interested in the same things." Do you think that's true? How do you pick your friends? Discuss what friendship means to you.
- By the time April's mother sends for her, April doesn't want to leave her grandmother or her new life. Why not? Have you ever wanted something very badly, but then when you got it, you realized you didn't want it anymore?
- What's special about the professor's Christmas gift to the children?

ABOUT THE AUTHOR: Zilpha Keatley Snyder was born in 1927 and raised in California, where her childhood experiences became the basis for most of her novels. The majority of her works are set in California. Ms. Snyder has said that she tries to write about "the kind of world we wake up to every day of our lives—but about a day that somehow turns out to be transformed into something strange and magical."

Beyond the Book...

HIEROGLYPHICS: In the book, the participants in the Egypt game take on Egyptian identities and develop their own hieroglyphics for their new names. There are books and kits about Egyptian hieroglyphics to share and discuss, and you can select a personal hieroglyph and explain why you chose it. Or create your own language, complete with an alphabet and punctuation.

MYTHOLOGY: Do some research on Egyptian mythology, gods and goddesses, and the Book of the Dead. Explain the characteristics and stories that surrounded the principal gods and goddesses, especially the ones that are important in the Egypt game. Someone else could research Egyptian archeology, including the pyramids, pharaohs, and Rosetta Stone, and explain how archeologists go about their work. Write your own myth based on what you've learned.

PAPER DOLLS: April and Melanie enjoy playing with paper dolls. Show each other any paper dolls that you may have (or make your own), and spend some time making up stories together about the dolls.

REFRESHMENTS OR FOOD MENTIONED IN THE BOOK: A favorite meal for the children is hot dog sandwiches and fruit salad, which may be served during your discussion of the book.

IF YOU LIKED THIS BOOK, TRY...
The Gypsy Game, by Zilpha Keatley Snyder—This is a companion novel.
Five Children and It; The Story of the Amulet; The Phoenix and the Carpet, by Edith Nesbit—These books also feature imaginative children who enjoy ancient history and other times.
Some Other Books by Zilpha Keatley Snyder:
The Headless Cupid
The Witches of Worm

Ella Enchanted

by Gail Carson Levine

Ella, as in *Cinderella* minus the "Cinder," is "blessed" at her birth by a well-meaning but inept fairy, who gives her the "gift"of obedience. She must always do as she's told. This entertaining novel follows the original story of *Cinderella,* with several clever embellishments, until our heroine breaks the fairy's spell by refusing to marry her beloved Prince, "Char."

As a mom, I secretly wish all three of my children were given the gift of obedience. Though she doesn't mind not having it herself, Morgan jokingly said she wishes her younger sister, Skylar, did—then she could really boss her around! But, as this story also tells us, we should be careful what we wish for.

READING TIME: 3–4 hours, about 232 pages
THEMES: obedience, independence, identity, women's rights

Discussion Questions

- What are the differences between this story and the original *Cinderella,* the fairy tale on which this story is based? Why do you think the author made each change?
- In most families, obedience is considered a good trait in children: A child who does what she or he is told is "good." In this story, obedience is considered a curse rather than a blessing. What is wrong with being too obedient? Do you think of yourself as obedient?
- Mandy is Ella's godmother, but she will do only "small magic," never "big magic." Why won't Mandy do "big magic"? Is there anyone in your life who helps you or protects you? Do you wish that your "fairy godmother" could do "big magic" and solve your problems for you? What would be wrong with that?
- When Ella's father asks her, "...who are you?" and then, "But who is Ella?" Ella's answer is, "A lass who doesn't wish to be interro-

gated." What is your answer to the question, "Who are you?"

- When Ella asks Hattie what would make her stop giving her orders, Hattie says, "If you stopped obeying them." Can you apply this dialogue to your own life? Perhaps you know someone who treats you in a way that you dislike. What could make the person stop? (Of course, Ella can't stop obeying orders, but you were not enchanted by a fairy!)
- When Lucinda asks Ella her name, Ella says, "Elle." That is her name in Ayorthaian, but in French, the word elle means "she." What do you think the author had in mind by having Ella call herself "Elle"? How might Ella's problem symbolize a problem of all women throughout history? How have modern women moved forward to help solve this problem? What problems still remain to be solved?
- In the end, who breaks Ella's spell? Could she have broken it all along? Why is she able to break it now?
- Of course, you're not under a spell or enchantment, but there may be something you would like to change about yourself, some way of acting or responding to others that would make you happier if you could stop or alter it. If you feel comfortable talking about it, discuss why such changes are so difficult to make and how you might make one.
- If you could give a baby girl a "fairy gift," what would you give her? Discuss your answers.
- What is the main idea the author wants to get across in this book?

ABOUT THE AUTHOR: Gail Carson Levine was a published poet before she graduated from high school. Since then she has continued to write and has even collaborated on a children's musical with her husband, David. She and her husband and their dog, Jake, live in an old farmhouse in Brewster, NY.

Beyond the Book...

FRACTURED FAIRY TALES: Think of another fairy tale you can retell to get an important idea across. Work in mother-daughter pairs or larger groups to come up with a new version of *Hansel and Gretel, Little Red Riding Hood,* or another traditional story. A good way to start would be to match up an idea with a story. Have group members act out your story, or parts of it.

WOMEN'S LIBERATION: In many ways, this book is about women's liberation. Many girls today take it for granted that women need not be obedient—that they have rights and can think for themselves. But many mothers may remember fighting for women's rights so that their daughters would not have to live under Ella's spell. Research the history of the women's rights movement. You may want to focus on a particular leader in the women's liberation movement. Share your findings.

REFRESHMENTS OR FOOD MENTIONED IN THE BOOK: There are plenty of treats in this book: butter cookies, plum pudding, currant bread, chocolate bonbons, spice cake with butter rum sauce, apricot sauce, and peppermint sauce. For a meal, make carrot soup, quail eggs, venison, or roasted pheasant stuffed with chestnuts. If someone's sick, make a realistic version of the curing soup, with carrots, leeks, and celery—leave out the hair from a unicorn's tail!

IF YOU LIKED THIS BOOK, TRY...

Half Magic, by Edward Eager—An ancient coin turns a dull summer into a series of magical adventures for four children (see p. 119).

Beauty, by Robin McKinley—This is a fairy tale retold from a feminist perspective.

Seven-Day Magic, by Edward Eager—Magic enters the lives of five children through a magic wishing book from the library.

The Woman in the Moon: And Other Tales of Forgotten Heroines, by James Riordan—In each of these 13 tales, the heroine is strong, bright, and clever.

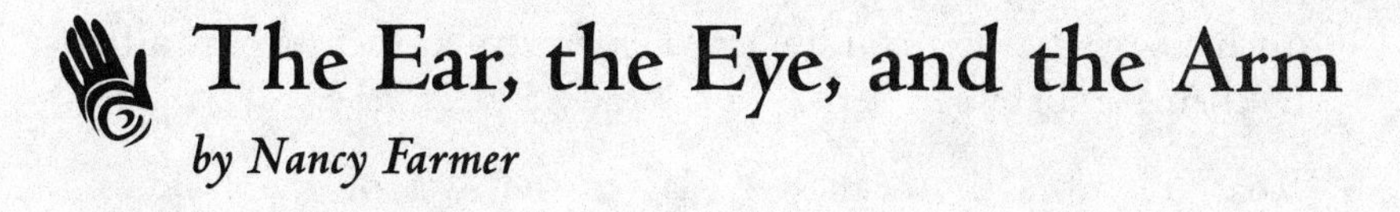

The Ear, the Eye, and the Arm

by Nancy Farmer

In this fast-paced, inventive novel set in 2194, the African country of Zimbabwe is overpopulated with displaced tribes who are ruled by military despots. The three children of a military chief long to escape their regulated lives and find out what Zimbabwe is really like. Within minutes of leaving home, they are kidnapped from the teeming urban streets that surround their fortress home. With warmth and humor, Nancy Farmer describes the children's contact with a messy, scary, fascinating population of poor, disenfranchised Africans, so different from their own family. In the end, three unusual detectives, the Ear, the Eye, and the Arm, help the children find their way back home. Along the way, the youngsters also discover their lost heritage and their most human qualities: bravery and compassion.

Of all the books our club has read, this is my favorite. In fact, I often use it in book club workshops, and each time, the discussion takes an unexpected turn and something new is pulled from the story. Morgan loves this book, too, especially the futuristic gadgets, the characters, and the adventure.

READING TIME: 4 hours, about 344 pages. I suggest that all readers preview the Appendix at the back of the book which describes the setting, tribes, language, and customs of the characters. This preview will be helpful in providing readers with the cultural context of the story.
THEMES: coming of age, courage, tribal heritage, wealth and poverty, gender, environmental pollution, independence, identity, compassion

Discussion Questions

✦ Describe the children's father. Do you think he is a loving parent? How do the children feel about him?

- ✦ In what ways has the children's father prevented them from fulfilling their Zimbabwean heritage of loyalty and bravery?
- ✦ What role does the Mellower play in the lives of the children's family? What does he give them that they cannot get for themselves?
- ✦ What are the Praise Songs that the Mellower sings about? Why are the songs so hypnotic and pleasurable? What memories do the songs arouse in the children's father?
- ✦ Why does Tendai free the myna bird early in the book?
- ✦ What are the children missing in their sheltered lives at home?
- ✦ Compare the children's well-protected mansion with Dead Man's Vlei where they are brought. How do the children feel about living there?
- ✦ What aspects of the messy "outside world" appeal to the children?
- ✦ The children's father regrets that he kept them from growing up. How does over-protection keep young people from reaching maturity? Do you think it's better to be over-protected or under-protected from the world beyond home and school?
- ✦ What role does the mother play in the family? Describe your impressions of the relationship between the mother and father.
- ✦ The children have a mixed reaction to the preserved tribal life of Resthaven. What aspects of that life do they love and hate?
- ✦ Describe the gender roles in Resthaven. How do Tendai and Rita feel about these roles? How are they similar to or different from gender roles in our society?
- ✦ The detectives, Ear, Eye, and Arm, possess exaggerated human qualities and skills. Describe their abilities.
- ✦ "As Arm said, too much Praise was bad for you, but a little was like a vitamin. It was necessary for healthy, happy spirits." What does this mean in terms of the book? Have you noticed this in your own life?
- ✦ By the end of the book, what have the children learned about themselves? Their country and their heritage?

ABOUT THE AUTHOR: Nancy Farmer did not begin writing books until the age of 40. Before that she lived in India, where she intended to teach English for the Peace Corps, but ended up teaching Chemistry instead. Through that position, she became a lab technician, ran a lab in Mozambique, and traveled to Zimbabwe where she met her husband, Harold. They have one son, Daniel, and now live in Menlo Park, CA.

Beyond the Book...

GEOGRAPHY: Take out your atlas and find some information about Zimbabwe. Talk about what conditions in modern-day Zimbabwe life might foreshadow the events and setting of the book. Look at some photos or slides of Zimbabwe.

READ ALOUD: Read aloud the descriptions of the detectives, the Ear, the Eye, and the Arm, in chapter 9 and on the first page of chapter 13. Imagine and draw pictures of the mutant detectives, using drawing paper and colored pencils.

GADGETS: Nancy Farmer has created some amazing futuristic gadgets and inventions. Think of some gadgets of your own. Draw what they would look like, then write up a few paragraphs about what they would do, how they would work, and who would use them. Share your ideas with each other.

IF YOU LIKED THIS BOOK, TRY...

The Giver, by Lois Lowry—This futuristic novel also portrays a young boy living in an overly-structured world and his longing for the diversity and chaos of lost human experience (see p. 97).

A Girl Named Disaster, by Nancy Farmer—Set in Zimbabwe, this painful coming-of-age novel follows the adventures of a young girl who must escape the limitations of her tribal life in order to fulfill her individual destiny (see p. 93).

Some Other Books by Nancy Farmer:

Do You Know Me

The Warm Place

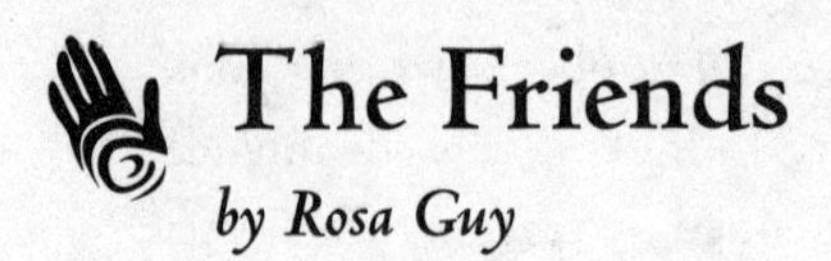

The Friends

by Rosa Guy

After life on a sunlit, West-Indies island, New York City seems cold, cruel, and dirty to 14-year-old Phyllisia Cathy. To make matters worse, her only friend, Edith, is unacceptable to Phyllisia's proud, tyrannical father because of her messy clothes and lack of refinement. The book ends with an uneasy truce between Phyllisia and her father, and a new ray of hope for Phyllisia's friendship with Edith.

After reading *The Friends,* our group invited Rosa Guy to be our first guest speaker at our meeting. During our talk, she told the girls that the story of *The Friends* was semiautobiographical, a fact that they found fascinating.

READING TIME: 3–4 hours, about 185 pages
THEMES: identity, rebellion, independence, friendship, economic status

Discussion Questions

- ✦ If you had a teacher who, like Miss Lass, was prejudiced against your racial or religious group, how would you handle her derogatory comments? Would you try to deal with this teacher on your own, with a group of classmates, with your parents, or a school administrator? What are the best ways of solving the problem of having such a teacher?
- ✦ Why is Phyllisia so unpopular among her classmates? Do you agree with her sister, Ruby, that Phyllisia should hide the fact that she is a good student so the other kids will not resent her? Should Phyllisia try to become more like them in order to be liked better?
- ✦ What do you think about the fact that Phyllisia uses her father's first name, Calvin, when she thinks about him? How do you think about your parents?
- ✦ Do you think Phyllisia's father loves her and her sister? What do you think about the way he does or does not express his feelings?

- ✦ Why was it so hard for Phyllisia's mother to communicate with her daughters? Did she prepare her daughters for her death? If so, how, and if not, why not?
- ✦ What was so hard about the way that Phyllisia's mother died? Have you ever known anyone who died of cancer?
- ✦ What does Phyllisia mean when she feels that Edith "doesn't go with the furniture"? Have you ever been in this situation with a friend? In discussion, find a way to handle this situation without creating a terrible, hurtful scene like the one that took place with Edith at Phyllisia's house.
- ✦ Put yourself in Edith's shoes. Imagine you have just left your friend's house after being insulted by her father, as Edith was by Calvin. How would you react?
- ✦ Which girl would you rather have as a friend, Marian or Edith? Discuss your reasons with each other. Also discuss why the author included Marian as a character.
- ✦ Think about Phyllisia's reaction when she finds that Calvin's restaurant is different from what she imagined. Why is she so disappointed? What insight into Phyllisia's character does her reaction give you?
- ✦ Reread Calvin's last words to Phyllisia and Ruby. Why does he scold them for having a messy room, when he threw all their clothes on the bed? What feelings is he hiding? Why does he feel he must hide his real feelings?

ABOUT THE AUTHOR: Rosa Guy has said that writers have a responsibility to try to make the world a better place. She brings this belief to her writing, which deals with world issues that she feels are crucial, such as race and gender. Ms. Guy was born in Trinidad and raised in Harlem, and has lived through many hardships that led her to become involved in the struggle for black freedom. She also helped to form the Harlem Writer's Guild. Ms. Guy has been writing for over 30 years and has won many awards for her young adult novels.

Beyond the Book...

POETRY WRITING: Using thoughts and ideas about friendship you developed from reading this book or your experience in a particular friendship, write a poem called "Friendship Is...," or "My Friend...." Begin each line with the two words of the title. Keep writing until you feel you have thoroughly defined what the word "friendship" or that particular relationship means to you.

ROLE-PLAYING: Each of you take the parts of Phyllisia and Edith and role-play a scene that might have occurred between them after Edith's visit to Phyllisia's house. You might try to play the scene in two different ways—one that ends with the two girls terminating their friendship, and one that ends with their friendship intact.

FAMILY DOS AND DON'TS: For daughters only. Start two columns on a sheet of paper labeled "Do" and "Don't." In the columns, list the rules that your parents want you to follow at home. Go back over your list and check the rules you think are fair or reasonable. Circle any you think are unfair or unreasonable. Openly discuss the fairness of your circled rules. Mothers and daughters may not end up agreeing, but at least you will get each other's thoughts on disputed topics.

MOVIE: Rosa Guy's goddaughter, Kathie Sandler, produced a short film version of *The Friends.*

REFRESHMENTS OR FOOD MENTIONED IN THE BOOK: For a meal or a side dish, make black-eyed peas. Or serve Mounds candy bars as a snack.

IF YOU LIKED THIS BOOK, TRY...

Ruby; Edith Jackson, by Rosa Guy—These two sequels to *The Friends* focus on Phyllisia's sister Ruby, and her friend Edith.

Nilda, by Nicholasa Mohr—A Puerto Rican girl grows from a child to a teenager in New York's *barrio.* In addition to a good story, Mohr provides profound insights into what it means to grow up poor and subject to prejudice.

The View from Saturday, by E. L. Konigsburg (see p. 272).

Some Other Books by Rosa Guy:

Disappearance

Billy the Great

From the Mixed-Up Files of Mrs. Basil E. Frankweiler

by E. L. Konigsburg

When two suburban children run away from home and take up residence in New York City's Metropolitan Museum of Art, they discover a side of the city—and themselves—that they didn't know existed. Their adventure has a surprising impact not only on themselves, but on the art world as well.

This is one of my favorite books. Maybe it's because I work at the Smithsonian Institution and have been in many museums before and after hours, or maybe I am just partial to smart, clever, resourceful heroines in action. I must also confess that I think it would be a lot of fun to spend the night in the Met. But for whatever reason, I love this story, and think it's fun and thought-provoking for girls and women alike.

READING TIME: 1–2 hours, about 159 pages
THEMES: family, secrecy, injustice, identity, self-esteem, confidence

Discussion Questions

- ✦ Why does Claudia choose to run away to New York City? Why does it evoke such strong feelings for her? Is there any place you feel as strongly about? Why?
- ✦ How does Claudia's attitude toward New York differ from that of her parents and her suburban community? What does that reveal about Claudia?
- ✦ Claudia is "bored with being straight-A Claudia Kincaid." Why? What does Claudia want from life? Have you ever been labeled, or felt like you had to meet other people's expectations of you?

- ✦ Why don't her parents pay attention to Claudia? And why don't Claudia and Jamie miss their parents when they're in the museum?
- ✦ Why does Claudia want to run away from home? Have you ever felt like doing that? Why?
- ✦ Why does Claudia want Jamie as her partner? How does she feel about her younger brothers, Steven and Kevin? Do you have siblings? How do you feel about them?
- ✦ Claudia says that she and Jamie complement each other perfectly. What does she mean? How do they become a team during their stay in the museum? Discuss the relationship they have, and how it changes during the book.
- ✦ Why does Claudia call running away a "great adventure"? What's your idea of a "great adventure"? Why?
- ✦ Why is the statue of the angel in the museum important to Claudia?
- ✦ Mrs. Basil E. Frankweiler observes that "often the search proves more profitable than the goal." What do you think she means by that? Have you ever experienced that?
- ✦ The search to discover the mystery of the angel makes Claudia, who is usually confident and self-assured, uncomfortable. Why? How does she feel about the new emotions that she experiences? Have you ever had a similar experience? How did it make you feel?
- ✦ Claudia cares about being different, and wants very much to be a "heroine." What does the word heroine mean to Claudia? Why is it important to her to be different from everyone else?
- ✦ What kind of relationship does Claudia have with Mrs. Frankweiler? How are they alike?
- ✦ Mrs. Frankweiler thinks that keeping her secret is an adventure—the kind of adventure that Claudia needs. Why? Are there any secrets you possess that you feel as strongly about as Claudia and Mrs. Frankweiler do?
- ✦ Mrs. Frankweiler says, "...some days you must learn a great deal. But you should also have days when you allow what is already in you to swell up inside of you until it touches everything and you can feel it inside you." What do you think she means?

ABOUT THE AUTHOR: Elaine Lobl (E. L.) Konigsburg was born in New York City, but spent the majority of her childhood in small mill towns in Pennsylvania. Growing up, reading was precious to her but it was less important to the rest of the family, who considered activities like "dusting furniture and baking cookies" more essential. In 1952 she received a degree from the Carnegie Institute of Technology (now Carnegie Mellon University), and went on to graduate school at the University of Pittsburgh. Before becoming a full-time writer, she married David Konigsburg and had three children, Paul, Laurie, and Ross. She worked for a while as a bookkeeper and then as a science teacher, but did not begin her writing career until her youngest child started school. She has written many well-loved, young-adult books, and her titles *From the Mixed-up Files of Mrs. Basil E. Frankweiler* and *The View from Saturday* have both won Newbery Medals (see p. 273).

Beyond the Book...

ART BOOKS: Take a day trip to the closest art museum. Choose one work of art you like there and research its history, the life story of the artist, and how the piece got to the museum in the first place. Try sketching it. Make a booklet about it, using a postcard or picture of the piece as the cover. Share your booklets with each other and explain why you like this particular piece of art.

BUDGET DAY TRIPS: Put yourselves on a small budget and then venture out for the day. Establish a treasurer who is in charge of the money. Take public transportation if it's available, and try to make your money last for three meals, as well as whatever activities you undertake during the day.

FILES: For fun, start a filing system for your personal collection of books or other treasures. Decorate the file folders and color code them: blue for mysteries, yellow for fantasy, and so on. Keep the files in a safe place.

ARTWORK: Draw, paint, or sculpt a piece that expresses something you feel deeply about. Display your work and explain how it represents those feelings.

GAME: Play the card game "War" like Jamie did with his friend on the bus.

MURAL: Check out a book on Michelangelo from your library. Familiarize yourself with his best-known works—the statue of David, and the painting on the ceiling of the Sistine Chapel. Paint a mural of angels to hang on the wall.

MOVIE: See the movie version of *The Mixed-up Files of Basil E. Frankweiler.*

REFRESHMENTS OR FOOD MENTIONED IN THE BOOK: Serve some of these goodies: cheese sandwiches, coffee, cereal, pineapple juice, peanut butter crackers, macaroni and cheese, baked beans, Hershey almond bars, hot fudge sundaes.

IF YOU LIKED THIS BOOK, TRY...

Harriet the Spy, by Louise Fitzhugh—A young aspiring writer plays detective among her friends and neighbors (see p. 122).

Some Other Books by E. L. Konigsburg:

The View from Saturday (see p. 272)

Jennifer, Hecate, Macbeth, William McKinley, and Me, Elizabeth

A Girl from Yamhill: A Memoir

by Beverly Cleary

In this bittersweet memoir, Beverly Cleary recounts the joys and pains of growing up as a beloved, but overly watched only child. The dashed dreams of her two parents struggling through the Depression contributed to a loving, practical, but demanding household. The joys of childhood, like an early farm life, wonderful times at her grandparents' general store, and playtime in a real neighborhood, are described in rich detail. So, too, are some of the woes of childhood: illness, difficult teachers, coping with caring but moody parents. Cleary transformed all of these experiences into memorable fictional form when she later became a writer.

READING TIME: 4 hours, about 344 pages
THEMES: education, independence, mother-daughter relationships, coming of age

Discussion Questions

- When times got tough, Beverly Cleary's mother gave her this advice: "Remember your pioneer ancestors." How did Cleary feel about this piece of advice? Do you think it had the effect her mother intended? Does your own family have a phrase like this to guide you through tough moments?
- To the young Beverly, growing up was filled with a lot of "nevers." Review the ones she mentions in the fifth chapter of the book. How do they affect her? Discuss some of the "nevers" in your own life.
- Beverly Cleary also says, "[My mother's] rules, if followed, would turn me into a little lady." Discuss some of these rules and what you think of them. Further, talk about whether there are spoken or unspoken rules today intended to turn girls into "little ladies."
- After a shaky start in early childhood, Beverly Cleary became an

avid reader, and she vividly describes the joys of her favorite childhood books. Talk about one or two of your own favorite books or stories you read or heard around the ages of six or seven.

- For her whole life, Beverly Cleary has remembered the sounds, sights, and smells of her first Portland neighborhood, which was filled with children. Bring your own childhood neighborhood to life by discussing some of its characteristics.

In reading this memoir, it's fun to see where some of the events in Beverly Cleary's novels come from.

- Beverly Cleary looked forward to her first day of school with great anticipation. Yet, sadly, the day was filled with confusion, rules, boredom, and embarrassments—harbingers of bad school days to come. Share recollections of your own first day of school.

- In Cleary's childhood, "...telephones were for the use of adults. Children who wanted to communicate with their friends stood on the front porch and yelled their names until they came out." By contrast, discuss the role of telephones, and even e-mail, in your own social life.

- Beverly suffered a number of humiliations at the hands of teachers in her early grades. These experiences pained her, yet she never lost a strong sense of who she was—not a nuisance, not short, and not slow. Where did her self-esteem come from?

- Cleary and her mother had a strong, loving bond, but a complicated one. Discuss their relationship. Are there any mother-daughter issues in the book you especially related to?

ABOUT THE AUTHOR: Beverly Cleary is known for composing her novels on a legal pad, and so may seem a bit old-fashioned in our modern, computerized world. However, her books tell us a different story. Because she was tired of the out-of-date themes in the books she read as a child, Ms. Cleary decided to base her characters on the children she grew up with. This decision produced amazing results, and her readers can identify with the characters she creates. She has written more than 30 books and has won many literary awards, including the Laura Ingalls Wilder Award and the Newbery Medal. Ms. Cleary and her husband, Clarence, live in Carmel, CA.

Beyond the Book...

OTHER BOOKS: If you have read some of Cleary's other books, discuss some of the parallels between her life and her work—experiences, themes, settings, characters.

READ ALOUD: In her memoir, Cleary recalls the significance in her life of the Greek myth of Demeter and Persephone. This myth tells the story of a powerful, often difficult, mother-daughter relationship. Track down this myth and read it aloud to each other. Follow up with a discussion of why this myth meant so much to Beverly Cleary and what application it might have to mothers and daughters in general.

MEMOIR: Cleary's memoir begins with vivid recollections of particular childhood scenes. Write a description of a memorable childhood event or scene to read to each other.

BOOK DRIVE: Books are a huge influence on Beverly Cleary's life, perhaps beginning when her mother single-handedly began a book drive to provide books for adults and children in Yamhill. Ask family members, friends, and neighbors to contribute books in good condition for donation to a local homeless shelter or day-care center.

REFRESHMENTS OR FOOD MENTIONED IN THE BOOK: The orchards Beverly Cleary describes in Oregon's Willamette Valley continue to provide our country with apples and cherries. As a nice tie-in with the early setting of the book, see if apples or Bing cherries from Oregon are available. Try some!

IF YOU LIKED THIS BOOK, TRY...

A reading or rereading of any Beverly Cleary book is the perfect follow-up to her wonderful memoir (see p. 205).

A Girl Named Disaster

by Nancy Farmer

This coming-of-age novel takes place in Mozambique where an unwanted girl, named Disaster by relatives who have reluctantly adopted her, sets out to find her long-lost father. Her dangerous journey takes her through jungles and infested waters. Disaster's intense love for her dead ancestors, along with her resourcefulness and a belief in herself that belies her unfortunate name, saves her from a disastrous arranged marriage and the limitations of her old village life.

READING TIME: 3–4 hours, about 300 pages
THEMES: identity, loneliness, courage, survival, death, redemption

Discussion Questions

✦ Why has Nhamo been given the name of Disaster? How does she deal with this label? What effect do you think unfortunate nicknames have on the people who are given them?

✦ From the beginning of the book, Nhamo seems to be searching for something: "...she didn't know what she wanted and so she had no idea how to find it." What is Nhamo searching for? Discuss ways in which this quotation might apply to your own life as well.

✦ Nhamo's dangerous journey tests her spiritual, emotional, and physical strength. Discuss what, if any, kinds of tests challenge today's young women.

✦ Every tribal girl in Nhamo's culture has a "bride price" that must be paid to the girl's future husband. How does Nhamo feel about this custom? What is your reaction to it?

✦ Several times, girls are described as being useless to the tribe. Since the tribal women are seen doing so much work, why do you think they are valued so little?

✦ Nhamo's well-regarded cousin Masvita seems to be the ideal young tribal woman: "She was modest and obedient. She never put herself forward, but kept her position respectfully equal with other people, and she did not say irritatingly clever things." In what ways does Nhamo deviate from this ideal? What do you think of this kind of ideal womanhood? Are some of these same qualities valued in today's culture as well? Why or why not?

- Describe how an individual's ancestors are regarded in the tribe. How do Nhamo's ancestors—her mother, grandmother, and those who lived before her—give her strength and courage? In what ways are your own ancestors remembered and valued in your family?
- The deceased relatives in Nhamo's past have as much influence and presence as those who are still alive. How is this a comfort to Nhamo? What does this tell you about how Nhamo's tribe regards death? How is it different or similar to the way death is regarded in our own society?
- In what ways do inner voices from your own relatives guide you through life?

Morgan loved this book because "it showed how independent girls can be and what they can accomplish if they put their minds to it."

- What do you think of this quotation: "Women are never free until they can control their own money."
- Nhamo comes into an inheritance from her grandmother, and the money is used for her education, something she loves and values. Nhamo gains so much in her new life, but do you think she has also lost something in moving away from her old way of life?

ABOUT THE AUTHOR: (see p. 81)

Beyond the Book...

CULTURAL CONTEXT: Review the nonfiction information in the back of the book describing the setting, tribes, language, and customs of the characters in *A Girl Named Disaster.* This review will be helpful in providing readers with the cultural context of the story. Talk about what equivalent or similar rites of passage or other customs there are in our culture.

VICTORY POEMS: Nhamo celebrates her bravery, survival, and ingenuity in her victory poems and songs. Try writing similar verses that celebrate your own personal achievements.

STORYTELLING: Nancy Farmer has woven a number of African folktales into the novel. As part of your book discussion, read some of these stories aloud since that is the way they were originally intended to be appreciated. For dramatic effect during the storytelling, dim the lights, burn candles, or light a fire if a fireplace is available. Afterwards, discuss these tales, what they reveal about the tribal belief system, what lessons Nhamo learns from them, and what hopes they give her.

WILD EDIBLES: Do some research on wild edibles, or visit a nature center where a naturalist can tell you more about them.

REFRESHMENTS OR FOOD MENTIONED IN THE BOOK: Corn is nutritious and easy to grow on African soil, and a staple in the diets of millions of Africans. Think of the many ways corn can be served—corn on the cob, cornbread, corn mush, polenta, corn tortillas, corn muffins, corn pudding, corn soup. Prepare and serve a few of these dishes at your book discussion.

IF YOU LIKED THIS BOOK, TRY...

Do You Know Me, by Nancy Farmer—In this novel, another young African girl must come to grips with her tribal identity and the developed world beyond it.

The Glory Field, by Walter Dean Myers—This novel follows five generations of one African-American family from Africa through the Civil War to contemporary America (see p. 100).

Julie of the Wolves, by Jean Craighead George—In this Alaskan novel, an Eskimo girl sets out in the wilderness to search for her father.

Some Other Books by Nancy Farmer:

The Eye, the Ear, and the Arm (see p. 79)

The Warm Place

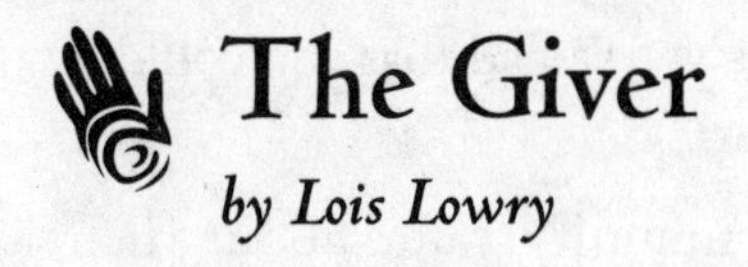

The Giver

by Lois Lowry

The Giver tells the story of 12-year-old Jonas who is chosen by his community to bear the weight of all human history. In Jonas's present world, there are few surprises. Children are tended with great care and assigned just the right names and roles to suit their personalities. Individuals' lives are programmed so pleasantly, with just enough quirks, that no one ever asks unsettling questions. The community has even planned for the transfer of human wisdom in the unlikely event that the lessons of the past are needed. Jonas sacrifices his ideal life in order to give to his community the pain, pleasure, color, and texture of human life.

READING TIME: 2–3 hours, about 180 pages
THEMES: freedom, individuality, choice, change, sacrifice

Discussion Questions

- What early signs show that Jonas questions the uneventful lives of his community more than others in his family and community do?
- When you first learned about the idea of "The Release," what did you envision? Did you develop any suspicions about what the release really was? What events in the book triggered your suspicions?
- Jonas's community has not quite banished its fears. What kinds of things make the community fearful?
- Why is The Giver unable to meet with Jonas every time?
- How are human feelings regarded and managed in Jonas's community?
- Why are knowledge and wisdom so exhausting? Do you think one person can embody all the sorrows of the world?
- What is the difference between talking about feelings and experi-

encing them? Can you describe feelings you've had that you couldn't completely express in words?

- ✦ What do the people of Jonas's community value about their planned lives? What do you think they miss?
- ✦ If you could live in Jonas's world, which parts of his life would you enjoy? Which parts would be difficult to experience?
- ✦ In what ways is our society like or different from Jonas's? Do you think today's society predetermines what people can become?
- ✦ Are there similarities between the use of pills in *The Giver* and our own use of medicines?
- ✦ What are the "stirrings" Jonas feels? Why do you suppose the community gives everyone a pill when they begin to experience these stirrings?
- ✦ There's an old saying: "Everyone complains about the weather, but nobody does anything about it." In *The Giver,* the community creates "perfect," unchanging weather. What have these humans gained and lost by living under ideal conditions all the time? If you could create endless, unchanging, perfect days, what kind of weather would you choose for those days? What would you miss about changeable weather and seasons?
- ✦ What color would you miss the most if you lived in a colorless world? How come?
- ✦ The most perfect memory The Giver passes on to Jonas is a festive, wonderful Christmas. If you have one perfect memory that you can give to someone as a present, what is it?
- ✦ What happens at the end of *The Giver?* Do they live or die? What does this ending mean?

ABOUT THE AUTHOR: A self-proclaimed bookworm, author Lois Lowry was always reading and writing as a child. Today her home is full of bookshelves that are filled to the brim with books. Once a photographer, Lowry always has clear pictures in her mind when she sits down to write. She has combined her writing and artistic talents many times, even supplying the photo for the cover of *The Giver.* Lowry enjoys reading, going to the movies, playing bridge, and gardening. She also likes to knit, and compares her writing to this craft. Lowry has said that composing a story involves piecing together patches of a quilt that is woven from scenes and dialogue. Lowry now lives in Cambridge, MA.

Beyond the Book...

ROUND ROBIN: In a round robin, pass around a notepad in which each book discussion guest writes one rule for a smooth-running society. Read the rules aloud and discuss what kind of society would result from following these rules.

RED APPLES: One of Jonas's first experiences with color is catching the flash of red of an apple. Fill a large bowl with perfect red apples to share during your book discussion of *The Giver.* Study the apples and speculate on why the author chose an apple as one of the first objects Jonas sees in color. Talk about the importance of colors in your own lives. What colors do you love? Hate? What associations do these favorites and nonfavorites have for you? If a person has never seen colors, what exactly is that person missing?

REFRESHMENTS OR FOOD MENTIONED IN THE BOOK: Make a meal of organic food and see if you can tell the difference. For dessert, serve freeze-dried ice cream. Compare it to real ice cream.

IF YOU LIKED THIS BOOK, TRY...

Escape to Witch Mountain, by Alexander Key—Two children search for their true home in the Smoky Mountains.

A Wrinkle in Time, by Madeleine L'Engle—A brother and sister search for their father in another time dimension.

Some Other Books by Lois Lowry:

All About Sam

Anastasia Krupnik

Autumn Street

Number the Stars (see p. 195)

The Glory Field

by Walter Dean Myers

This sweeping family saga celebrates the African-American experience. It tells the story of the Lewis family, from Muhammad Bilal, who is brought to South Carolina in chains in 1753, to the 15-year-old Malcolm Lewis, who, in 1994, is headed for a future filled with endless possibilities.

This book made me think of my own family, and how small I thought it was until I attended a 50-person family reunion about five years ago. This reunion made me realize how easy it is to lose touch with one's family—and how important it is not to.

READING TIME: 4–5 hours, about 240 pages
THEMES: race, human rights, family, responsibility, courage

Discussion Questions

- ✦ Describe how slavery is portrayed in this book. How do other portrayals you have read or seen of slavery compare with this one? How does reading or seeing such portrayals in books, movies, or on television make you feel?
- ✦ In "April 1900" Grandma Saran asks, "Why can't they leave our men alone!" Grandpa Moses answers, "The only way some people can see their own manhood is by pushing somebody else down." Do you agree with his explanation? What reasons might you add? Is her statement only true about manhood or about people in general?
- ✦ In "March 1864" Miss Julia invites Lizzy to tea. In "May 1930" Florenz Deets and her friend Katie invite Luvenia Lewis to tea, then use her, unthinkingly, for their own purpose. Compare the two episodes in as many ways as you can. Why do you think the author included two such similar episodes?
- ✦ In "January 1964" Mr. Chase shares his opinions about integration with Tommy Lewis. What are Mr. Chase's thoughts on the

subject? Suppose someone expressed the same ideas to you. What would your reaction be? Would you argue against them? How would you try to refute them?

- ✦ What motivates Tommy to risk his scholarship by publicly confronting the sheriff? Suppose you were Tommy's friend. What would you have advised him to do—stay away from the meeting or get involved? Give your reasons.
- ✦ In "August 1995" Jenn Che Po refers to Malcolm's earlier words to her: "You know, what you said about starting with being yourself, who you are, and then moving on?" What do these words really mean? Can you apply them to yourself? If so, how?
- ✦ Compare these parallel scenes: Malcolm and Shep are riding in the stifling back of the truck, where they gasp for air through a grate; back in 1753, when the men "forced their way beneath the hatch opening so that they could suck in the occasional breeze." Why do you think the author draws these parallels between the lives of characters from earlier and later generations?
- ✦ What is the importance of the family reunion to Malcolm? How does it enrich his life?
- ✦ Why is holding on to the land so important to the characters in this book? What is the significance of "the land"? What role does it play in their lives?
- ✦ Why do you think it was worth more than $200 to Planter Lewis to buy back the shackles that had been worn by his ancestor, and later by Tommy Lewis? Why do you think he left them to Malcolm?
- ✦ In "March 1864" Lizzy has her "freedom dream." The dream theme is continued throughout the book: In "April 1900" Elijah dreams of an even larger reward than the one he received from Mr. Turner. In "May 1930" Luvenia says, "Curry isn't my dream, it's my daddy's," and in "August 1994" Malcolm dreams of being successful with his instrumental group, "String Theory." In which ways are all these characters' dreams "freedom dreams"? What is your dream? Is it a "freedom dream" as well?

ABOUT THE AUTHOR: Walter Dean Myers was inspired to write *The Glory Field* by what he saw as a "change in the texture" of the African-American community from generation to generation. Mr. Myers was born and raised in Harlem, NY, and went on to write many well-loved novels for young-adult readers. He has not only won awards for his fiction writing, but for his nonfiction and poetry as well. He now lives with his family in Jersey City, NJ.

Beyond the Book. . .

FAMILY TREE: Look over the Lewis family tree at the beginning of the book. Then make your own family tree, asking older relatives to help you, if necessary. When you have completed your family tree, share it with other members of your family. Talk with them about your "roots," and what being aware of your family heritage means to you.

ORAL HISTORY: The young people of the Lewis family are reminded of their family history and traditions by members of the older generations. Collect stories and traditions that your family has handed down from one generation to the next. You might tape record or videotape interviews with family members and later transcribe them onto the pages of a family history book.

FREEDOM SONGS: In this book, the author refers to the traditional African-American freedom songs, "Sooner in the Morning When I Rise," and later, "O Freedom!" Research in the music section of the library to find and learn these songs, as well as others, such as the famous Civil Rights movement song, "We Shall Overcome." Get together and sing these songs with family members or friends. Then talk about the words and their meaning.

FAMILY REUNION: Encourage your family to have a family reunion and help plan or host it.

REFRESHMENTS OR FOOD MENTIONED IN THE BOOK: While I was growing up, my mom made the best fried chicken, and we would have Sunday dinners every week that were made up of a lot of the foods mentioned in this book. Make your own Sunday dinner with fried chicken, collard greens, rhubarb pie, sweet potatoes, rice, and hot bread pie. Follow it up with some sassafras tea.

IF YOU LIKED THIS BOOK, TRY...

Malcolm X: By Any Means Necessary, by Walter Dean Myers—This is a biography.

The House of Dies Drear, by Virginia Hamilton—A history professor and his son investigate their rented house, formerly a station on the Underground Railroad, unlocking the secrets and dangers from attitudes dating back to the Civil War.

Some Other Books by Walter Dean Myers:

Now Is Your Time

Somewhere in the Darkness (see p. 235)

Scorpions

Fallen Angels

The Young Landlords

The Gold Cadillac

by Mildred D. Taylor

Two young girls painfully learn the meaning of segregation when they journey with their parents from their sheltered home in Ohio into Mississippi. The lessons they absorb on this frightening trip remain with them long after they return home, and forever change the way they see the world.

Growing up, I listened to my mom and her girlfriends, whom I called aunts, tell stories about segregation. I thought I knew what the word meant. Many years later, I attended an exhibition at the Smithsonian Museum of American History where I had to choose between "White Only" and "Colored Only" doors to exit the museum. Only then did I truly experience what segregation meant. It is an experience, not just a word, that provides true understanding. Watching Wilma and 'lois go through the same realization in this book gives readers an idea of what that experience is like.

READING TIME: 1–2 hours, about 30 pages
THEMES: race, discrimination, family, values, courage, self-worth

Discussion Questions

- Wilma and 'lois's father is proud of his new gold Cadillac. How does their mother react? Why do you think she responds that way? Has there ever been a surprise that has had unanticipated reactions in your family?
- What does the car mean to their father? Why does it have such significance?
- What kind of things does the girls' mother think are important? What about their father? How are their values different?
- Why wasn't it a good idea for their father to drive the car down to Mississippi?

- ✦ Why do you think the girls' mother changes her mind about the car, and agrees to travel down to Mississippi?
- ✦ One of the girls' uncles says, "we might've fought a war to free people overseas, but we're not free here!" What does he mean? What must that feel like?
- ✦ When the family crosses the Ohio River into Kentucky, their father tells the girls not to speak to white people. Why?
- ✦ When her relatives were first packing for the trip down South, Wilma and 'lois thought the preparations were fun. How does their own trip to Mississippi make them see those efforts differently? How would you have felt in that situation?
- ✦ How does 'lois feel when the police take her father to jail, and don't believe the car is his? What message does that send? Have you ever received similar messages in the way people treat you?
- ✦ Even though the Cadillac is important to her father, after the trip to Mississippi, he sells it. Why?

ABOUT THE AUTHOR: Novelist Mildred Taylor was determined to become a writer by the time she was in high school. Born in Jackson, MS, she grew up in Toledo, OH, graduated from the University of Toledo, and taught English and history in the Peace Corps in Ethiopia. She spent her childhood listening to true stories of slavery and the African-American experience. In addition to *The Gold Cadillac* and Newbery Medal winner *Roll of Thunder, Hear My Cry,* Ms. Taylor wrote the Coretta Scott King Award winner, *The Friendship,* inspired by her family's stories.

Beyond the Book...

JIM CROW: Before the group meets, have someone research and prepare a short presentation on the Jim Crow laws, and how segregation affected African-Americans in the South. Someone else can draw a map explaining the migration of southern African-Americans to the industrial North, a massive exodus that had a profound impact on family relationships and family structures.

PRE-CIVIL WAR: This story takes place in the pre-Civil Rights era. Discuss who Rosa Parks was, and the importance of the Montgomery bus boycott and the Woolworth counter sit-ins. This will provide some historical context for this family's experiences with segregation.

REFRESHMENTS OR FOOD MENTIONED IN THE BOOK: Wilma and 'lois's family had to prepare food for the trip because they wouldn't be able to eat at any of the restaurants along the way. Have a picnic featuring the foods they ate during the trip, like fried chicken, sweet potato pie, and potato salad, and talk about what it would have meant not to have choices about being able to eat on the road. Talk about how things are different today—and whether they are actually as different as they are supposed to be.

IF YOU LIKED THIS BOOK, TRY...

Roll of Thunder, Hear My Cry, by Mildred Taylor—Prejudice and racism prevail under the Jim Crow system (see p. 212).

A Raisin in the Sun, by Lorraine Hansberry—She explores the realities of life for African-Americans in the North.

Some Other Books by Mildred D. Taylor:

The Friendship

The Golden Compass

by Philip Pullman

When Lyra Belacqua is called upon to save the world from destruction in this unusual fantasy, she little suspects that this quest will force her to confront unspeakable and disturbing mysteries at the center of her own life. The choices she makes and the causes she allies herself with will ultimately alter her own destiny forever.

This is a must-read for anyone with a sense of adventure and curiosity. I was really intrigued by the whole concept of daemons, which seem to serve a similar function as blankies or stuffed animals, as well as the discussion of destiny.

READING TIME: 4–5 hours, about 351 pages
THEMES: love, family, values, trust, betrayal, choice, identity, courage

Discussion Questions

- ✦ What kind of relationship does Lyra have with her uncle? How do her feelings for him change during the course of the book?
- ✦ Lyra has had a somewhat unconventional upbringing, being raised by the professors and scholars at a college. How has that upbringing influenced her idea of "family"? What does "family" mean to you?
- ✦ Why is Roger's friendship important to Lyra? Do you have a relationship that's special to you in the same way?
- ✦ How does Lyra feel when she is told that she has to leave the college? What kind of relationship does she have with Mrs. Coulter, and how does that change during the course of the book?
- ✦ What role does her daemon, Pantalmaion, perform in her life? Is there someone or something that plays that role in your life?
- ✦ The witches have heard about Lyra's peculiar destiny. What does it mean when their consul observes, "But she must fulfill this destiny

in ignorance of what she is doing, because only in her ignorance can we be saved." What do you feel your destiny is?

- ✦ Loyalties are constantly shifting in this book, as Lyra pursues her quest. How does Lyra decide who to trust, and when? How does she decide what to believe? Have you ever trusted someone who turned out to be untrustworthy or believed in something that turned out to be false? What was that experience like? How did it affect you?

- ✦ How does her relationship with the armored bear, Iorek Byrnison, influence the course of Lyra's destiny? What's significant about their bond?

- ✦ What qualities in her personality help Lyra succeed in her quest? How do the traits that might otherwise be considered faults end up helping her?

- ✦ People see the same events or phenomena differently, depending on who they are in this book. For example, while flying seems foreign and extraneous to most of us, the witches say they could no more stop flying than stop breathing. Is there something so essential to who you are that you couldn't imagine living without it? Discuss what that is, and why.

- ✦ Serafina Pekkala tells Lyra, "You cannot change what you are, only what you do." What do you think she means by that? Do you agree or disagree? Are there things about "what you do" that you'd like to change?

- ✦ The witches live longer than any of the mortals they may love. What do they lose, and what do they gain, by their longevity? Would you want to be in their situation? Why or why not?

- ✦ What does it mean when Lord Asriel betrays Lyra?

ABOUT THE AUTHOR: Philip Pullman spent his childhood in a variety of places—from Africa to Australia to England and Wales. He attended Oxford University where he studied English literature. He went on to become a teacher and now divides his time between writing and lecturing at Westminster College, a division of Oxford University.

Beyond the Book...

FANTASY WORLD: One element of fantasy is constructing another world. Come up with your own imaginative universe, complete with maps and underlying philosophy, and share the results when you discuss the book.

BIBLE: A central theme of this book has to do with the Bible, and interpretations of what original sin means. Read the story of Adam and Eve, and find appropriate resources and other stories to give an idea of what the doctrine of original sin has come to mean in Western European culture.

REFRESHMENTS OR FOOD MENTIONED IN THE BOOK: Serve spice cakes and chamomile tea as refreshment, which were treats for Lyra in the book.

IF YOU LIKED THIS BOOK, TRY...

The Subtle Knife, by Philip Pullman—Lyra continues on to further adventures.

The Hobbit; Lord of the Rings, by J. R. R. Tolkien—This fantasy series chronicles the Hobbit's adventures.

Pawn of Prophecy, by David Eddings—This is the first book in a fantasy series.

The Chronicles of Narnia, a series by C. S. Lewis—English children explore a magical land.

A Wrinkle in Time, by Madeleine L'Engle—A comparably feisty and intrepid heroine is undaunted in her rescue efforts.

Some Other Books by Philip Pullman:

The Ruby in the Smoke
Spring-Heeled Jack
The Tin Princess

Good Night, Mr. Tom

by Michelle Magorian

An abused child of a single mother, Willie Beech is sent to the English countryside during World War II to escape the London bombings. In the country he discovers friendship and family that he has never known before. This brief idyll, paradoxical though it is, offers Willie the opportunity to transform his life.

READING TIME: 2–3 hours, about 318 pages. Be prepared for some difficulties with the regional English country dialect.
THEMES: friendship, trust, abuse, loyalty, grief, redemption, cruelty

Discussion Questions

- How does Willie see the world when he first goes to live with Tom Oakley? Why?
- How does Willie's initial treatment of Tom's dog, Sassy, reflect how Willie himself has been treated during his life?
- Tom's life closed in on him after his wife, Rachel, died while giving birth to their son. How does Willie's arrival change Tom's life?
- Willie's mother has seen him as a nuisance at best, and evil at worst. He observes that "For her to like him he had to make himself invisible." What does Willie mean by that? Discuss how that affects his self-image and behavior.
- Why is Zack's friendship important to Willie? Have you ever had a friend who's been that important to you?
- When Willie is returned to his mother, he feels diminished—"his own mother made him feel ill." Have you ever known someone who made you feel that way? How do you imagine it would affect you if it were your own mother?
- After Tom brings Willie back from London, beaten and bruised, the doctor says, "The sores will heal. They healed before. It's the

wounds inside that will take the longest to heal." What does he mean? What internal wounds does Willie have?

- ✦ Zach and Will are "evacuees," children who have been relocated to the English countryside for safety during World War II. In what ways are their lives and the lives of the other village inhabitants affected by the war, even as they are removed from raids and immediate danger?

- ✦ How have Will and Tom changed each other? Have you been changed by a relationship with someone? How so?

ABOUT THE AUTHOR: Michelle Magorian's parents met during World War II and she was born in Portsmouth, England, in 1943. The war serves as a backdrop for many of her books. Ms. Magorian spent part of her childhood in Singapore and Australia, but returned to England when she was nine years old, at which time she began to write stories. She became an actress and a poet as a young adult. Ms. Magorian is married and has two children.

Beyond the Book...

LIFE DURING WORLD WAR II: Find out if any members of the group have parents or grandparents who remember living through World War II, and see if they can share memories of how their daily lives changed because of the war. Research what life was like for American children here, where the war wasn't being fought but they were nonetheless affected by it. Have someone go to the library to research oral history accounts of English children who, like Willie, had been sent to live in the countryside. Discuss what it would be like to have to leave your home and parents to go live with strangers.

PLAY: Stage and produce a pantomime/pageant like the one that Willie and his friends participate in. Imagine what it was like to live in a time and place where children, and adults, were largely responsible for their own entertainment. What do you think it would be like to live without television, Nintendo, or other amusements? Would you like it? Why or why not? What do you think children gained or lost by living in that kind of society?

REFRESHMENTS OR FOOD MENTIONED IN THE BOOK: Unlike today, when any foods can be easily obtained at the local supermarket, people living in the English countryside during the war had limited options. What we consider ordinary snacks, like lemonade or chocolate cake, were special treats for children then. Prepare a special, festive tea, complete with scones, butter, and ginger cake to share during the discussion.

IF YOU LIKED THIS BOOK, TRY...

The Secret Garden, by Frances Hodgson Burnett—This is a similar tale of a spurned and rejected child who gets a renewed chance in the English countryside.

The Kingdom of the Sea, by Robert Westall—An English boy's family is killed in the bombing of Britain during World War II.

Autumn Street, by Lois Lowry—This book describes life in a small town during the war years.

The Summer of My German Soldier, by Bette Greene—This tells the story of a friendship between a former prisoner and a young Jewish girl (see p. 244).

Some Other Books by Michelle Magorian:

Back Home

In Deep Water and Other Stories

The Great Gilly Hopkins

by Katherine Paterson

Tough-talking, troublemaking Gilly Hopkins has been shunted from so many foster homes she's developed a thick shell that keeps her own feelings in and keeps everyone else's out. What she doesn't anticipate, though, is meeting Trotter in her final foster home. Trotter's unwavering patience and love helps Gilly develop the self-worth that's been buried inside her.

READING TIME: 2 hours, about 148 pages
THEMES: rejection, alcoholism, loss, self-worth, identity, personal survival

Discussion Questions

- ✦ Gilly Hopkins speaks and behaves in one way, but the author gives us a glimpse into her inner self. How does Gilly Hopkins appear to other people? How does she appear to you?
- ✦ Why is Gilly so difficult and tough toward everyone she meets?
- ✦ Gilly meets her match in several people. Who are the characters that refuse to respond to Gilly's orneriness? What is it about them that breaks through Gilly's tough, protective shell?
- ✦ Gilly steals, she can be mean, she picks on people's most sensitive flaws, and at times she's even racially prejudiced. Still, she has redeeming qualities. What are they?
- ✦ Every time the word "mother" is mentioned, Gilly gets a terrible feeling inside. Why is that? Describe Gilly's feelings for her mother.
- ✦ What is it about William Ernst that gets through to Gilly?
- ✦ What role does Mr. Randolph play in the story? What insights do you gain about Mrs. Trotter from the relationship with Mr. Randolph?
- ✦ Gilly's foster mother Trotter is full of wise sayings. What do you think this one means? "Once the tugboat takes you out to the

ocean liner, you got to get all the way on board. Can't straddle both decks." What was your favorite of her sayings?

- Gilly must leave the one foster home where she wants to remain. However, she goes off to her grandmother's house with a very different attitude than the one she had about her previous foster homes. What has caused Gilly's change of heart?
- Gilly finally gets to meet her mother. Compare the image Gilly has of her mother with the actual person.
- How does Gilly respond to her real mother?
- Gilly's grandmother wants to get to know her by talking face-to-face. Gilly reacts to this approach with these thoughts: "You can't talk it out, you got to live into their lives, bad and good." What does this mean? Do you agree with Gilly's view about how people must get to know each other?
- Compare the way Gilly was at the beginning of the book to the way she is by the end. What changed her?

ABOUT THE AUTHOR: (see p. 23)

Beyond the Book...

VOLUNTEER: You might want to organize a project to help homeless children by raising money for local charities that serve the homeless population. Some areas of the country collect money to buy brand-new back-to-school clothes for homeless or poor children who may never have had the experience of wearing brand-new clothes purchased just for them. Brainstorm ways to raise funds to buy new clothes for several needy children in the area.

READ ALOUD: Obtain a copy of *The Oxford Book of English Verse* and read aloud some of the poems mentioned in *The Great Gilly Hopkins*. Discuss why these poems meant so much to the characters. Find a poem that means something to you or that you like, read it aloud, and explain why you feel this way about it.

REFRESHMENTS OR FOOD MENTIONED IN THE BOOK: Gilly Hopkins just loves the food that her beloved Trotter prepares for her. Plan a fried chicken dinner with mashed potatoes, cranberry salad mold, and cherry pie. Mothers and daughters may prepare this meal on a potluck basis. Hold the *Gilly Hopkins* book discussion around the dinner table. Talk about why lovingly prepared food tastes so good! What does Trotter's meal represent for Gilly? What kinds of family feelings do people experience when they prepare and enjoy a meal together?

IF YOU LIKED THIS BOOK, TRY...

Ellen Foster, by Kaye Gibbons—An 11-year old girl struggles to find herself as she copes with her mother's death and the neglect and violence of her remaining family.

Thursday's Child, by Noel Streatfield—A tough ten-year-old, orphaned girl in Victorian England builds a defensive wall around herself to protect herself from her feelings of pain and loss.

Some Other Books by Katherine Paterson:

Bridge to Terabithia (see p. 21)

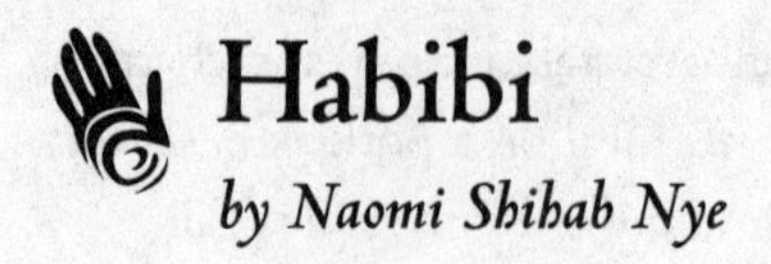

Habibi

by Naomi Shihab Nye

Political conflicts affect the personal life of 14-year-old Liyana Abboud, a half-American, half-Arab girl who moves from Missouri to Jerusalem at a pivotal moment in her life. The discoveries she makes about herself, her family, and her new homeland help her attain a surer sense of who she is, and who she wants to become.

Before reading this book, I always focused on the Arab-Israeli conflict from the Israeli point of view, mainly because that's what I was exposed to most. *Habibi* allowed me to see this conflict from another perspective.

READING TIME: 2–3 hours, about 259 pages
THEMES: friendship, values, first love, family, identity, prejudice

Discussion Questions

- ✦ How does Liyana feel about moving from St. Louis to Jerusalem? Why? How would you feel if your parents were about to make such a move?
- ✦ Liyana is only able to take some of her possessions with her on the move. What does she take? If you were moving halfway around the world, what would you take with you if your choice was limited the way Liyana's is? Why would you make these selections?
- ✦ Liyana's father has always talked about Palestine as his home. When they arrive at the Tel Aviv airport in Israel, her family is singled out because they are Arabs. How does that affect her father? How does that make the rest of the family feel?
- ✦ Her younger brother, Rafik, observes, "I thought when I got to the other side of the world, I might feel like somebody different." What do you think he means? Does he feel different in Jerusalem?
- ✦ How do Liyana and Rafik initially respond to their father's mother, Sitti? How do their feelings change during the course of the book? How do they feel when they see the West Bank village?

- ✦ Liyana discovers that she, along with the rest of her Arab family, carries the burden of history with her. For example, her father is unwilling to take them into Jewish Jerusalem because of the long-standing conflict between the Jews and Arabs. How does this affect Liyana?

- ✦ There is also tension between Liyana's American upbringing, and the new world of Arab tradition and culture that she now is a part of. How does this change the way Liyana deals with her parents?

- ✦ When her brother, Rafik, finally feels happy living in Jerusalem, Liyana feels even more alone. Why is that? Have you ever been in a situation where someone else's comfort caused you discomfort? What was that like?

- ✦ Living on the West Bank, how does Liyana redefine her idea of friendship?

- ✦ Her grandmother, Sitti, is full of superstitions. Share superstitions that you may have gotten from your grandmother (or that your mother got from hers). Are there any that cross cultural boundaries? Are there any that are the same?

- ✦ Why is Liyana's relationship with the Jewish Israeli boy, Omer, important? How does it change during the course of the book?

- ✦ Liyana says, "When you liked somebody, you wanted to trade the best things you knew. You liked them not only for themselves, but for the parts of you that they brought out." What is she referring to? Are there people who make you feel like that? Who are they, and what parts of you do they bring out?

- ✦ How does the destruction of Sitti's bathroom by an Israeli soldier change how Liyana and her family feel about the country? How does her father's arrest change how Liyana feels about the political situation?

- ✦ The book says of Liyana that "she had always felt homesick for some other life. . . ." What life was she longing for? Have you ever felt this way?

ABOUT THE AUTHOR: Naomi Shihab Nye has written several books for young adults and is also a prolific poet and essayist. She recently published a collection of essays for young adults called *Never in a Hurry,* which the *School Library Journal* claims is "...guaranteed to make [young adults] laugh, cry, reflect, or think about life from another point of view." Ms. Nye lives in San Antonio, TX, with her husband, Michael, and their son, Madison.

Beyond the Book...

JEWISH-ARAB HISTORY: Do some research into the history of Jewish-Arab relations, including the Partition of Palestine, the Six-Day War, the Lebanese conflict, and the recent Intifidah, to gain a better understanding of the political landscape of the book. Find a map of the Old City of Jerusalem, and locate the neighborhoods and different quarters where Liyana and her friends spend their time.

ORAL HISTORY: As part of her school work in the United States, Liyana is supposed to do an oral history with an older relative. Try to do a brief oral history with a grandmother, great-aunt, or other older woman, and share these stories with the group.

REFRESHMENTS OR FOOD MENTIONED IN THE BOOK: To convey the atmosphere of the book, serve Middle Eastern dishes like yogurt, falafel, humus, baba ghanoush, dates, apricots, and oranges.

IF YOU LIKED THIS BOOK, TRY...

One More River; Broken Bridge, by Lynne Reid Banks—Each deals with the Arab-Israeli conflict, and the relationships between Jewish and Arab teenagers.

Some Other Books by Naomi Shihab Nye:

Benito's Dream Bottle

Lullaby Raft

Half Magic

by Edward Eager

Four children are thrilled when they discover a magical coin that can transport them anywhere in space or time. The twist is that the magical token grants them only half of what they wish for, resulting in lots of amusing mishaps along the way. As a result, an otherwise dull summer is transformed into a spirited adventure that ultimately changes their lives.

READING TIME: 1–2 hours, about 192 pages
THEMES: imagination, family, responsibility

Discussion Questions

✦ The book starts off by stating that Katherine is a comfort to her mother. In general, what kind of relationships do the children in the book have with the adults in their lives? Are they like the relationships you have with the adults in your life?

✦ How much of the book is based on historical settings or events? Specifically, what settings are recreated?

✦ How is their mother's parenting style different from other mothers? Would you want to have their mother as your own? Why or why not?

✦ How do the children react to discovering an ancient coin that turns out to have magical powers? If you stumbled upon that kind of magic, what would you wish for? Why?

✦ Some of the first wishes made on the coin don't have quite the results that the children expect. How did Mark resolve the problem? What was it about the way he phrased the wishes that made them effective? What would you have done?

✦ When the children meet Merlin the wizard, he cautions them against "terrible good intentions." What do you suppose he means by that? Have you ever been in a situation where you meant well, but somehow everything turned out wrong? How did that make you feel, and what did you do about it?

- Why does Jane react to Mr. Smith so differently than her siblings do? What finally changes her mind about him?
- Why does the charm come back to grant Jane's last wish? How has the charm made the family's lives better? What about Jane's in particular?

ABOUT THE AUTHOR: Edward Eager, born in 1911, began writing children's books in 1951 as a result of his search for stories to read to his son, Fritz. He worked as a playwright and lyricist as well as a writer, and died in 1964 at the age of 53.

Beyond the Book. . .

KING ARTHUR'S COURT: Have some members of the group prepare a presentation on Camelot and King Arthur's court to stimulate discussion of the children's adventures. Bring in books on tournaments, jousts, and chivalry, and ask the girls to imagine what it would have been like to live in such a world. What would have been appealing about it, and what would have been difficult? What do they think it means to live in a world where behavior followed such strict codes?

PICNIC: Plan a lakeside picnic that includes food from the book like deviled eggs and potato salad. Talk about what it would have been like to live in a simpler time when children had more freedom to pursue such activities unsupervised by adults.

MOVIE: Rent the video *A Connecticut Yankee in King Arthur's Court* and have a discussion about how that character's experience in Camelot was similar to, and different from, the children's adventures. Talk about the possibility of time travel. If you could go anywhere in history, where would you go? Why? How would you avoid changing history because of your knowledge about what happens?

REFRESHMENTS OR FOOD MENTIONED IN THE BOOK: One of the biggest treats for the children in this book was going downtown to the local soda fountain for ice cream desserts. Plan your own "soda fountain" party, and replicate the ice cream concoctions from the book or create your own.

IF YOU LIKED THIS BOOK, TRY...

Bedknobs and Broomsticks, by Mary Norton—A fantasy that offers a similar type of fantasy/adventure that includes time travel.

Five Children and It, by Edith Nesbit—This book was the inspiration for *Half Magic.*

Some Other Books by Edward Eager:

Knight's Castle

Magic by the Lake

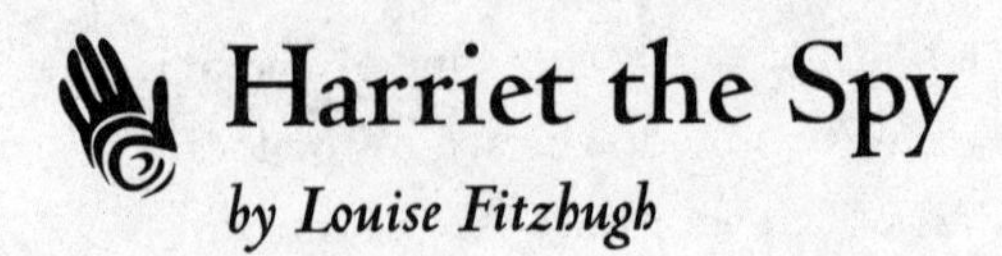

Harriet the Spy

by Louise Fitzhugh

Harriet, a plucky sixth-grade girl, is determined to become a writer someday. Her curiosity and ambition lead her to spy and take detailed notes on the activities of classmates, neighbors, and friends. Her secret life as a spy is exposed when her classmates accidentally discover her notebook.

Harriet's baby-sitter, Ole Golly, is one of my favorite characters. My children had the same baby-sitter for 15 years, Greta. She arrived each morning at 7:30 and left each evening at 6:30 like clockwork, and missed only two days of work in 15 years. She was an integral part of our family. Reading this book, I couldn't help but compare Harriet's experience with my children's.

READING TIME: 2–3 hours, about 150 pages
THEMES: coming of age, privacy, secrecy, separation, friendship

Discussion Questions

- ✦ What do you think of Harriet's spying on everyone?
- ✦ Why does Harriet have such a strong curiosity about other people?
- ✦ Do you think spying is an effective way to find out what people are really like?
- ✦ Have you ever overheard people say things that really surprised you?
- ✦ What would you do if you learned that a classmate or acquaintance was spying on you or had secret negative opinions of you?
- ✦ Harriet keeps close tabs on her caretaker, Ole Golly, but Ole Golly totally surprises her anyway. What was Ole Golly up to that caught Harriet completely off guard?
- ✦ Ole Golly is full of wise observations that she shares with Harriet. What are some of the values that she passes on to Harriet?

- ✦ Have you ever had a strong bond with someone who took care of you besides one of your parents? If you have had such a relationship, describe some of the activities you shared with that person and some of the values that person passed on to you.
- ✦ Compare the relationship Harriet has with her caretaker, Ole Golly, to her relationship with her mother and father.
- ✦ Ole Golly quotes the writer Dostoyevsky to Harriet: "If you love everything, you will perceive the divine mystery in things." What do you suppose this means?
- ✦ Harriet asks herself: "Is it terrible to get what you want?" Do you think it might be terrible to get what you want? Describe something you wanted and received that wound up disappointing you.
- ✦ Despite all her detailed note-taking on everyone's activities, Harriet realizes people can still take her by surprise. She asks herself: "Is everybody a different person when they are with somebody else?" Discuss whether or not you think this is true.

ABOUT THE AUTHOR: Louise Fitzhugh did not set out to be a writer. She first hoped to be an artist, a talent she employed later in her career to illustrate her own books (including *Harriet the Spy*). Her writing career began when she collaborated on a book called *Suzuki Beane* with her friend Sandra Scoppetone. Following this collaboration, Fitzhugh tried to write on her own. The process of writing did not come easily to her, but she was willing to work very hard, and with the help of her editors and friends, she created what many have called a breakthrough book. The realistic view of parents and children in this book broke the tradition of presenting a world that adults wanted children to see, not the world in which children actually lived. Fitzhugh died at age 46.

Beyond the Book...

WRITING: Take notes on snippets of conversations overheard in public places. Note details about these speakers' gestures, clothes, and speech. Create a story or imagined world based on these observations.

GAME: The Telephone game is a good springboard for discussing the perils of passing on gossip or making assumptions based on partial or misheard information. Isn't this what happens to Harriet? Telephone is initiated when the first person in a round robin writes down a sentence then whispers it to the next person who passes it on. By the end of the round robin, the sentence is usually a hilarious, mangled version of the original.

MOVIE: See the movie version of *Harriet the Spy.*

REFRESHMENTS OR FOOD MENTIONED IN THE BOOK: Harriet's daily after-school snack is milk and cake. Share your favorite after-school and after-work snacks during the book discussion.

IF YOU LIKED THIS BOOK, TRY…

The Long Secret, by Louise Fitzhugh—This is a sequel to *Harriet the Spy.*

Anastasia Krupnik, by Lois Lowry—This is the first in a series of books about another outspoken female heroine who creates her own family and school adventures.

The Great Gilly Hopkins, by Katherine Patterson (see p. 113).

Some Other Books by Louise Fitzhugh:

Nobody's Family Is Going to Change

A Hero Ain't Nothin' but a Sandwich

by Alice Childress

Told from several characters' points of view, this realistic novel of a poor, inner-city family portrays the life of promising 13-year-old Benjie, whose experimentation with heroin causes his parents and teachers to intervene. Benjie, his parents, friends, and teachers offer a loving but candid portrait of Benjie's descent into drug addiction.

READING TIME: 2–3 hours, about 128 pages. The explicit language and volatile issues of this powerful novel are for mature readers only.
THEMES: drug abuse, race, poverty, family

Discussion Questions

✦ When the book opens, Benjie Johnson says: "Now I am 13, but when I was a child, it was hard to be a chile. . . ." Benjie makes his childhood sound as if it took place a very long time ago. Why did childhood end so soon for Benjie? When do you think childhood ends?

✦ Benjie also says: "Don't nobody wanta be no chile cause, for some reason, it just hold you back in a lotta ways. . . ." Why does Benjie think it's better not to be a child any longer? What does it mean to be a child?

✦ Benjie can't stand people lying to him, but he tells a lot of lies to himself. What are some of the lies he tells himself about the people who care about him and about his drug problem?

✦ What impressions do you get from Benjie about his mother? About Butler? His grandmother? Jimmy Lee? Mr. Cohen? Nigeria Greene? How do your impressions change when they speak in their own voices about themselves and their feelings for Benjie?

✦ How do the people in Benjie's life feel about him, and how does Benjie feel about them?

- ✦ What makes Benjie an appealing character to everyone despite his toughness and his way of keeping people at a distance?
- ✦ Why do you think Benjie shuts out everyone around him?
- ✦ Butler says: "Poor folks ain't got many ways to solve problems." In what specific ways would more money have helped the family's situation and Benjie?
- ✦ Many of the characters who speak about Benjie's drug problem have tough things to say about social workers and psychologists. Describe how Benjie, Butler, Nigeria Greene, and some of the other characters feel about the professional people who make a living from the problems of poor people. Do you agree or disagree with some of their opinions about the officials in the "helping professions" like psychologists and social workers?
- ✦ Benjie's teachers, Nigeria Greene and Bernard Cohen, disagree about everything but Benjie, yet they actually have a lot of personality traits and beliefs in common. Discuss what they disagree about, then discuss surprising ways they are similar. In the end, what is each man trying to accomplish?
- ✦ Nigeria Green and Bernard Cohen are on opposite sides of a racial divide. Compare their racial views and discuss where you agree and disagree with each of them.
- ✦ Benjie's friend, Jimmy Lee, has a hard time watching Benjie destroy himself with drugs. Describe what it's like for Jimmy Lee to see his friend hurting himself so much.
- ✦ Why do you think Benjie takes drugs? What do you think of Benjie's excuse that he isn't really addicted. Do they think Nigeria Greene and Bernard Cohen were right to turn Benjie in?
- ✦ Benjie's mother wants so much to say to him: "Benjie, the greatest thing in the world is to love someone and they love you too." But what comes out of her mouth is: "Benjie, brush the crumbs off your jacket." Discuss your thoughts on why she can't really say the words of love she wants to say.
- ✦ What excuses does Walter, the drug pusher, make for selling drugs to kids? How did you react to his views about his role in drug pushing?

- What do you think Benjie means when he says to his stepfather: "A hero ain't nothing but a sandwich"? Discuss whether you agree or disagree with this statement.
- When the book ends, do you believe Benjie is about to meet his stepfather or not?

ABOUT THE AUTHOR: Alice Childress was a woman of many talents. Actress, novelist, playwright, lecturer, and director are only some of the words that describe her. Ms. Childress earned the OBIE award, making her the first woman ever to win one. She was also the first black woman to have a play produced on Broadway. She died of cancer in 1994 at the age of 77.

The multiple points of view in this classic make it unusually honest and intimate.

Beyond the Book...

DRUG DISCUSSION: Studies have shown that the children of parents who share their views about drugs are less likely to experiment with drugs. *A Hero Ain't Nothin' but a Sandwich* presents an opportunity for mothers and daughters to have a discussion about drugs, using incidents from the book as a starting point. Mothers can point out that although the setting of the book is the inner city, drugs are a problem everywhere. Can daughters imagine the same situation taking place in a suburban area as well? If the discussion is going well, the mothers may want to offer information to help girls figure out positive ways to handle the pressures that lead to drug experimentation.

RESEARCH: Do some research on how drugs affect one's system and why they are so dangerous. Invite an expert to speak at the group meeting, or go as a group to a discussion on drugs at a school, church, or community center.

REFRESHMENTS OR FOOD MENTIONED IN THE BOOK: Serve fried chicken during your discussion. Have you ever had hand-dipped ice cream? If not, go get some. If your talk about the book is in the morning, make bacon, pancakes, and coffee cake.

IF YOU LIKED THIS BOOK, TRY...

The Friendship, by Mildred Taylor—A young boy refuses to call a white man "Mister" and gets his family in trouble.

The Ups and Downs of Carl Davis III, by Rosa Guy—A black boy has a hard time adjusting to small-town life after his family sends him away to South Carolina to straighten out.

Some Other Books by Alice Childress:

Rainbow Jordan

Those Other People

Homecoming

by Cynthia Voigt

James, Sammy, Maybeth, and Dicey have been abandoned by their emotionally disturbed mother. Now Dicey must either find their mother or a new home where the family can stay together. After a long trek to a great-aunt's house, the four children are disappointed. Finally, to everyone's surprise, they find a loving home with their eccentric grandmother.

READING TIME: 4–5 hours, about 372 pages

THEMES: coming of age, family, separation, responsibility, independence, hardship

Discussion Questions

- Instead of going to the authorities for help, Dicey feels she must find her mother's aunt Cilla. Why does she feel this way? Do you agree with her choice? What would you do in Dicey's situation?
- What qualities does Dicey have or need to take care of her family and get them to their destination successfully? Can you think of other situations in life that require similar character traits?
- Louis, the boy the children meet camping at the beach, tells them that children have no rights. Do you agree this is true? Which children's rights would you support?
- The kids sang a lot in this book. Why was the music important to them?
- Louis says, "Look out for yourself and let the rest go hang—because they're out to hang you. . . ." Do you agree with this outlook on life? If not, what advice would you give the children instead?
- After James steals $20 from Stewart, James says it doesn't matter because "Nothing matters. There's nothing you can count on—

except the speed of light. And dying." Do you agree? If not, how would you convince James that being honest—or just being a good person—is important, even though everyone dies in the end?

- Dicey's definition of "home" is "where you rested content and never wanted to go anywhere else." What is your definition of "home"?

- What is your opinion of Cousin Eunice? Explain why you like or dislike her.

- Dicey says, "Maybe life was like a sea, and all the people were like boats." Then she tries to imagine what kind of boat she would be. She wonders if every boat (person) needs a harbor, or if some can just sail with the wind all their lives. What do you think? What kind of boat would you be? Do you need a "harbor"?

- Jerry, the boy who gives Dicey and the children a ride on his sailboat, says to Dicey, "Rebellion is necessary for the development of character." What does he mean by that? Do you agree? Why, or why not?

- Of all the people Dicey and the children meet on their travels (Louis and Edie, Windy and Stewart, Will), who do you like the best, and why?

- Why doesn't Gram want the children to stay with her permanently? What makes her change her mind?

ABOUT THE AUTHOR: Cynthia Voigt is the author of numerous books for young adults. Her writing career began when she finally found time to pursue her writing between her teaching career and family. Ms. Voigt's novels are based on real-life occurrences: Often just a fragment or image she sees prompts her imagination to finish the story that is playing out in front of her. She has won the Newbery Medal and other awards for her contribution to children's literature.

Beyond the Book...

CAMPING: Dicey and the children have to travel on foot. They sleep, cook, and eat outdoors while they travel. If possible, camp out for a night, bringing with you as little equipment as possible. Imagine that you have no home to go back to. Around the campfire, talk about how Dicey and her family must have felt traveling as they did. Sing some songs, and see if it has the same effect on you that it has on the characters in the book.

FOSTER CARE: Find out as much as you can about the laws in your state protecting abandoned children. The Internet offers a wealth of information and a wide range of views on foster care. Use "Foster Care" as your search words. After pooling your research, talk about what would have happened to Dicey and her brothers and sister if they had been placed in foster homes.

MAPS: Dicey depends on her maps to get her family where they need to go. On road maps of Massachusetts and Connecticut, try tracing the actual journey Dicey might have taken from Peewauket to Cousin Eunice's house in Bridgeport. Using the scale of miles, figure out how many miles they traveled on foot to New Haven, and then by car to Bridgeport. Then, on a road map of your town, find a place not too far away you would like to visit. Trace the route on the map. You might extend the same activity to a car trip. Adults do the driving (of course), and children navigate, using a road map.

REFRESHMENTS OR FOOD MENTIONED IN THE BOOK: You might try cooking some of the same food the children ate—potatoes, fish, chicken, fruit.

IF YOU LIKED THIS BOOK, TRY...

Dicey's Song, by Cynthia Voigt—This is the sequel to Homecoming.

Where the Lilies Bloom, by Vera and Bill Cleaver—The smart, gutsy heroine of this novel is determined to fulfill her promise to her dying sharecropper father—to keep the family together (see p. 289).

Some Other Books by Cynthia Voigt:

Bad, Badder, Baddest

David and Jonathan

Homesick: My Own Story

by Jean Fritz

Jean Fritz, born and raised in China until the age of 12, portrays those exciting growing-up years in this fictionalized account of her experiences. A coming-of-age story, this novel details the day-to-day life of an expatriate family living in the midst of a bustling China in 1959, at the time of the Communist revolution.

READING TIME: 3 hours, about 163 pages
THEMES: homesickness, patriotism, family ties, coming of age, homecoming, living abroad

Discussion Questions

- On the opening page of the book, Jean Fritz sets these two sentences, one after the other: *"Twenty-five fluffy little yellow chicks hatched from our eggs today, my grandmother wrote. I wrote my grandmother that I had watched a Chinese magician swallow three yards of fire."* Why do you think the author mentions these incidents side by side so very early in the book? How do these sentences get you ready for the rest of the book?
- Jean Fritz has been quoted as saying: "No one is as eager to find roots as the person who has been uprooted." What do you think this quotation means? Discuss incidents in the book that exemplify the meaning of this observation. If you have ever been away from your own home for a length of time, discuss how this quotation could apply to your own experience.
- Jean Guttery, the heroine of this book, has strong feelings about being an American. What experiences deepen her sense of patriotism while she is in China?
- From the descriptions in this book, what do you know about life in China at the time Jean Fritz was there? What do you know about life in America at the same time? How were they different?
- Why can't Jean sing "God Save the King" at her British school?

- ✦ Jean Guttery is always opinionated and often outspoken in her opinions. What does she have strong opinions about? What do you have strong opinions about?

- ✦ One of Jean's outspoken opinions regards her name, which she doesn't believe suits her at all: "The name Jean was so short, there didn't seem to be enough room in it for all the things I wanted to do, all the ways I wanted to be." What are some of the things she wants to be? Jean Fritz chose her very own first name as the first name of her heroine. What does this tell you about her adult feelings regarding her name? Do you think your own name has "enough room in it" for all the things you want to be? Discuss other names, if any, that you wish you were named instead.

- ✦ Jean says: "Deep in my heart I knew that goodness didn't come natural to me. If I had to choose, I would rather be clever, but I didn't understand why anyone had to choose." Discuss whether you would prefer to be good or clever. Do you think people can be both?

- ✦ The minister at Jean's church compares death to a trip he'd once taken when "...the train went through a long dark tunnel. Then suddenly it burst out of the tunnel into a blaze of light and you were in Italy. That's what death was like, he said. It was a glory. Nothing to feel sad about." What do you think of this image of death? Do you have any visual images of death?

- ✦ How important was it for Jean to communicate with her grandmother in the United States?

- ✦ Jean is outwardly respectful towards most adults she encounters, yet she says: "But part of me was never sure about grown-ups." Daughters: What do you think the heroine means by this? Do you agree or disagree? If you agree with that view, what kinds of things make you unsure about grown-ups? Mothers: Do you remember ever feeling unsure about grown-ups in a similar way? How does adulthood confirm or belie that point of view?

- ✦ Jean Guttery's 12 years are spent in much longing for America. Yet, when she actually sees China slipping away, what are her feelings about leaving it behind?

- ✦ Life in America isn't totally perfect for Jean Guttery. In what ways is it as she had always hoped? What disappointments does she experience?

ABOUT THE AUTHOR: Jean Fritz often does her best writing while the rest of us are sound asleep. Though she usually writes all day long, she finds that things seem to come together at night when she is trying to sleep. Ms. Fritz writes by hand, and is constantly rewriting. She finds that once she has an idea, she cannot stop writing until she is finished. Ms. Fritz said that she does not need to find ideas for books, they find her. She believes that books are made up of outside influences as well as what you give of yourself. Her own experiences have often provided material for her books. An avid reader, Jean Fritz recommends that young people who enjoy reading and might like to write should keep a journal so that they can be in touch with their emotions, something she feels is essential for a writer. Ms. Fritz has written numerous books, including a number of biographies. She currently lives in Dobbs Ferry, NY.

Beyond the Book...

PHOTOS: Prior to the book discussion, ask any members who have lived or traveled overseas to bring in photographs or letters to share with the group. Invite them to compare their own experiences overseas with those of the characters in Jean Fritz's book.

MAP: Have a globe, world atlas, and U. S. map on hand so that you can trace the Gutterys' journey from Hankow China (now called Hangzhou) all the way to Washington, PA. As you trace the route, discuss Jean's feelings at each step of the journey—pulling away from China, the ocean voyage, arrival in San Francisco, the car trip across the United States, and Jean's final homecoming to her grandmother's house.

CHINESE SOCIETY: In 1959, China was on the brink of big changes. Go to the library (or ask a history teacher for help) and find some information about that period to enhance your enjoyment of Jean Fritz's work.

REFRESHMENTS OR FOOD MENTIONED IN THE BOOK: Morgan and I were the hosts when our club read this book. We had a great discussion, but the food stole the day. We served some great Chinese food, and topped it off with a Castle Cake just like the one described in the book, with Chinese writing and all!

IF YOU LIKED THIS BOOK, TRY...

China Homecoming, by Jean Fritz—Jean Fritz wrote an account of her return to China in 1982 after the publication of *Homesick: My Own Story.*

A Little Princess, by Frances Hodgson Burnett—One of Jean's favorite fictional heroines is Sara Crew, the main character of this classic novel.

Young Fu of the Upper Yangtze, by Elizabeth Foreman Lewis—China's famous river makes a deep impression on Jean Fritz's heroine. This novel takes place along the Yangtze River during pre-revolutionary China where a boy is apprenticed to a coppersmith. A 1932 Newbery Medal book.

Some Other Books by Jean Fritz:

Brady
The Buffalo Knife
The Cabin Faced West
Bully for You, Teddy Roosevelt

The House of Dies Drear

by Virginia Hamilton

This suspenseful mystery tells the story of a black professor and his family who move into a huge old house once inhabited by an abolitionist. The family soon discovers that the house, full of secret passages and underground tunnels, was a stop on the Underground Railroad. Present and past racial conflicts intersect when some townspeople begin to harass the new family as soon as they arrive. However, the father and one of his sons figure out a dramatic and haunting plan to end the harassment.

Virginia Hamilton is one of my favorite young-adult authors. She really respects her readers' intelligence. Her novels are always multilayered and full of intrigue.

READING TIME: 3–4 hours, about 279 pages
THEMES: slavery, prejudice, adventure, mystery, friendship, trust

Discussion Questions

- Why does Thomas's father want so badly to live in the house of Dies Drear? How does the rest of the family feel about it?
- When do you start to realize that there is something strange and haunted about the house the Smalls are moving into?
- Describe Mr. Pluto. What is odd about him? Whose side do you first believe he's on?
- Have you ever lived in or near a house that people considered to be haunted? Do you believe in ghosts? If so, explain what makes you think they exist. If not, then discuss why you think other people believe in them.
- How does history repeat itself when the Smalls move into the house?

- ✦ Why did Mr. Pluto and Thomas trust Pesty? Describe the basis for the friendship between Thomas and Pesty.
- ✦ Do you think Pesty was in conflict with her own feelings? Why or why not?
- ✦ Describe the relationship between Pluto and his son. Why do you think it was so important for Mr. Pluto to remain independent and live by himself?
- ✦ Bias incidents, like the Los Angeles riots and recent church burnings, continue to plague communities across the country. Talk about these kinds of bias incidents: their causes, the effects on people who suffer from them, and the prejudices that engender them.
- ✦ The Smalls figure out a dramatic and clever way to end the Darrows' harassment of them. Describe the events they stage to frighten these people away. Do you believe Thomas and his father were justified in teaching the Darrows a lesson?

ABOUT THE AUTHOR: (see p. 52)

Beyond the Book. . .

UNDERGROUND RAILROAD: Before reading *The House of Dies Drear,* make copies of information about the Underground Railroad to share. This will provide a helpful context for the past events portrayed in the book. Talk about the Underground Railroad as part of the discussion of this book.

GROUP HIDEAWAY: Our group read this book during our first year, and the mother-daughter team who hosted the meeting went all out. The meeting reminder was designed like a treasure map, with directions to their house. For the meeting, they turned their basement into a maze, leading to a small, hidden room set up like a stop on the Underground Railroad. We were led to the secret room by clues similar to those in the book. After the book discussion, the girls returned to the room and ate on floor palettes as if they were journeying on the Underground Railroad. It was extremely rewarding to watch the book come to life at our meeting.

IF YOU LIKED THIS BOOK, TRY...

The Mystery of Drear House: The Conclusion of the Dies Drear Chronicle, by Virginia Hamilton—This is a sequel to *The House of Dies Drear.*

The True Confessions of Charlotte Doyle, by Avi—This is another mystery featuring a heroine caught up in historical events (see p. 266).

Some Other Books by Virginia Hamilton:

Arilla Sun Down

Drylongso

Zeely

Plain City

Sweet Whispers, Brother Rush

The Hundred Dresses

by Eleanor Estes

The painful effect of children's casual cruelty is the subject of this powerful book about peer rejection. A school clique is insensitive to a bashful, shabbily dressed newcomer. When the girl's family is forced to move again, her imaginative legacy, exquisite drawings of 100 dresses she imagined but never owned, reminds her classmates of their own role in her disappearance.

A simple and timeless lesson for us all!

READING TIME: 1 hour, about 80 pages
THEMES: popularity, teasing, tolerance, economic status

Discussion Questions

- In what ways is Wanda Petronski portrayed as an outsider?
- What is your impression of Peggy? Of Maddie?
- Describe how the teasing of Wanda Petronski begins. What keeps it going? Do you think there is anything Wanda might have done to stop the girls from teasing her?
- At first Maddie has a little fun at Wanda's expense, but it soon begins to bother her. What prevents Maddie from stopping the teasing or speaking against it?
- Describe Peggy and Maddie's friendship. What do you think of their relationship?
- The author says: "Peggy was not really cruel. She protected small children from bullies. And she cried for hours if she saw an animal mistreated." Do you think it's possible for a person who isn't basically cruel to engage in teasing anyway?
- What was your first reaction when Wanda announced that she had 100 dresses and 60 pairs of shoes at home?

- In what way does the teasing start to ruin not only Wanda's life but Maddie's as well?

- What do you think of the idea of school uniforms? Would they have made a difference for Wanda? Would you like to have school uniforms? Why or why not?

- Do you think Wanda created the drawings of the 100 dresses before the teasing began or afterwards?

- Throughout the book, Maddie is full of good intentions that she doesn't fulfill. What are some of her good intentions? What would have happened if she had acted on her intentions instead of just thinking about them?

- What do you think of the letter Peggy and Maddie write to Wanda after she wins the art contest?

- Do you think Maddie agrees with Peggy's statement: "This shows she really liked us. It shows she got our letter and this is her way of saying that everything's all right." Does Maddie agree that "everything is all right" with Wanda? Do you?

ABOUT THE AUTHOR: When Eleanor Estes was growing up she was surrounded by open fields, trees to climb, and perhaps most important, stories. Her mother was a dramatic storyteller whose stories were laden with details and images that would come to life before her children's eyes. Ms. Estes credits this environment for her development as a writer. The line between imagination and reality was often blurred for Ms. Estes; her books are filled with her childhood imaginings and embellished memories, which to her were often indistinguishable from what actually occurred. Eleanor Estes loved being with people and led a full life until her death in 1988.

Beyond the Book...

CHARITY: Organize a clothing drive in your book discussion group or at school. Collect favorite articles of clothing you would like to donate to a charitable organization.

CLOTHING DESIGN: Bring in a photograph or a drawing of a favorite outfit and talk about your special memories of wearing that outfit. Design and draw a beautiful outfit for someone you know. Share your drawing at the book discussion for *The Hundred Dresses.* Or have daughters design outfits for their mothers and vice versa.

PAPER DOLLS: Create your own paper doll set. Make a figure of a girl out of cardboard, and have each member of the group draw paper dresses to fit her. Talk about why dressing up dolls—paper and real—is such a popular activity for girls growing up.

SCHOOL UNIFORM: Design the ultimate school uniform. Share your idea and explain why you designed the way you did.

REFRESHMENTS OR FOOD MENTIONED IN THE BOOK: Prepare a favorite school lunch—brown bag or lunchbox style—and share it during your discussion.

IF YOU LIKED THIS BOOK, TRY…

Anything for a Friend, by Ellen Conford—A lonely girl, desperate for friends, thinks up various schemes to impress everyone.

The Empty Schoolhouse, by Natalie Savage Carlson—A young girl experiences rejection when she is the first black girl to attend her school.

Invisible Lissa, by Natalie Honeycutt—Fifth-grader Lissa, tired of being an outsider, challenges the popular, powerful girls in her class.

Some Other Books by Eleanor Estes:

Ginger Pye

The Curious Adventures of Jinny McGee

I Know Why the Caged Bird Sings

by Maya Angelou

In this poignant autobiography, the author's evocative account of her experiences as a young black girl growing up in the racist, rural South during the Depression of the '30s and early '40s, as well as in wartime San Francisco, offers insight into the condition of the black female during the pre-Civil Rights era. She forges her own identity, despite encountering a multitude of obstacles, and gains control of her destiny.

I have never been so moved by a book as I was by this one. The African-American experience is rich; it is filled with pain and oppression, but there is also incredible love and joy. I thank Maya Angelou for laying bare her soul and a part of all of us.

I have had the pleasure of meeting Maya Angelou and introducing her to Morgan. After chatting with us for about five minutes, she turned to me and said, "Thirteen is not a country." When I looked confused, she explained that it was from a poem I needed to read. Her office faxed it to me the next day. The name of the poem is *Portrait of Girl with Comic Book,* by Phyllis McGinley, and it's a wonderful description of the in-between feeling of being thirteen, when one is still a young girl but trying to be a lady.

READING TIME: 3–4 hours, about 289 pages. Some of the material, which includes descriptive scenes of sexual abuse, rape, and violence, is best suited to more mature readers.
THEMES: self-esteem, pride, love, family, segregation, relationships, race, prejudice, trust, abuse, values, self-discovery, hope

Discussion Questions

- What's special about the relationship between Marguerite and her brother (who calls her Maya)? Why is it that way? If you have siblings, how do you feel about them? Do you have more of a special bond with one than with the others?

- At the ages of three and four, Marguerite and her brother, Bailey, are sent to live in Arkansas with their paternal grandmother, Momma. How do you think they felt? Imagine making that kind of trip at that age. How do you think it would have affected how you grew up?
- Momma has definite expectations for Marguerite's behavior—with other adults, in church, or when she helps out in the family store. Do your parents expect you to conform to a certain code of behavior? If so, where and how does it apply? How is your code different from the one Marguerite has to follow? Why do you think Marguerite's behavior was so important to her grandmother?
- What kind of relationship does Marguerite have with Momma? How is Marguerite's view of the world influenced by Momma?
- Even though her grandmother owns land and houses, Momma's life—and Marguerite's—is still profoundly affected by the pervasive racism in Arkansas at the time. The black and white communities are almost separate worlds. How does Marguerite see the white world?
- What kind of relationship does Marguerite have with her parents?
- When she is seven, Marguerite rejoins her mother for a few months, but she feels as if she's with a stranger. She observes, "But what mother and daughter understand each other, or even have the sympathy for each other's lack of understanding?" What does she mean? Why does she feel that way with her mother? Have you ever felt that way with yours?
- How does Marguerite feel at her eighth-grade graduation when she and her classmates are told to aspire only to athletics? How would you feel if someone limited your dreams and aspirations?
- At the age of ten, Marguerite goes to work for a white woman. What lessons does she learn from her, both intentional and unintentional?
- Why is Mrs. Flowers important to Marguerite? Have you ever known an adult who had that kind of influence or impact on you? Marguerite says that "[Mrs. Flowers] had given me her secret word which called forth a djinn who was to serve me all my life: books."

What does she mean by that? Has anyone ever opened your eyes to something like that?

- Have you ever felt ignored, disrespected, or as if your comments were of no value? Can you imagine how Marguerite felt about the blatant disrespect the white people showed her? Did Marguerite and her grandmother handle their feelings differently? How so?
- Why was the church and Marguerite's belief in God so important in her life? How did religion help to shape her values?

ABOUT THE AUTHOR: Maya Angelou was born in 1928 in a rural town near St. Louis, AR. Her given name was Marguerite Johnson. Her young life was not easy, but her accomplishments are extraordinary. She began her career in drama and dance and is now considered an author, poet, playwright, musician, historian, actress, producer, director, and Civil Rights activist. She has made landmark achievements in music (by becoming San Francisco's first black female conductor), film (with Emmy-award nominations for her performances in *Roots* and *Georgia*), and literature (with both Pulitzer Prize and National Book Award nominations). *I Know Why the Caged Bird Sings* is the true account of her childhood. Ms. Angelou, who speaks six languages, is currently lecturing around the country and has spent some time as a professor of American Studies at Wake Forest University, in North Carolina.

Beyond the Book...

OBSERVING: Maya stopped talking for five years and became a keen observer of everything around her. See if you can go a whole day without talking. What do you notice about the world around you? How do people react to you? What happens to your other senses?

METAPHORS: Maya Angelou is the consummate poet and storyteller. This book is filled with wonderful metaphors. Find some good examples of metaphors from the book, and talk about your favorites, and why metaphors in general can be so powerful.

FAITH: Religion and church are major influences in the African-

American culture portrayed in *I Know Why the Caged Bird Sings.* Attend services at a church with an African-American congregation or listen to gospel music to get a sense of the pervasive role such expressions of faith have in that community.

MAP: Before your discussion, prepare a map showing the migrations of blacks from the rural south to the industrial North, similar to the journey Marguerite's parents took, and explain what kind of effect such moves placed on family life. Talk about what it was like for husbands and wives to be separated from each other, or for children to be separated from their parents.

REFRESHMENTS OR FOOD MENTIONED IN THE BOOK: The annual community fish fry/picnic was a major event. Prepare your own celebratory feast, including lemonade, fried chicken, sweet potato pie, potato salad, caramel and coconut cakes. Or serve Hershey's Kisses™ and canned pineapple slices for refreshments; talk about how these plentiful items, that are easily found in supermarkets, were such significant treats during the Depression.

IF YOU LIKED THIS BOOK, TRY...

A Raisin in the Sun, by Lorraine Hansberry—A play about a different element of the black experience.

The Bluest Eye, by Toni Morrison—For older readers, this book deals with race and self-image. It would be a good choice for exploring this issue further.

Some Other Books by Maya Angelou:

The Heart of a Woman
Wouldn't Take Nothing for My Journey Now
I Shall Not Be Moved

In the Year of the Boar and Jackie Robinson

by Bette Bao Lord

In 1947, Shirley Temple Wong sails from China to her new home in Brooklyn, NY. Shirley finds America to be a land of wonders, but she doesn't know any English, nor does she understand many American customs, so it's hard to make friends. Finally, Shirley's indomitable spirit wins her friends, and listening to radio broadcasts of the Brooklyn Dodgers' fight for the pennant helps her become fluent in English.

READING TIME: 2–3 hours, 169 pages
THEMES: difference, cultural identity, tolerance, perseverance, family

Discussion Questions

- ✦ What can you learn about China from the first few drawings in the book? Have you had experiences or known anyone who has had to learn the language and customs of a new country or culture like Shirley Temple Wong does in this book? If not, perhaps you've read a book or seen a movie about someone who does. Drawing on what you know about such experiences, discuss what is most difficult for kids like Shirley. What advice would give a girl like Shirley to help her make friends and be happy?
- ✦ How would you help a girl like Shirley who came to your school from another country and culture?
- ✦ Joseph tries to befriend Shirley by inviting her to play stoopball. Why does Shirley have to leave the game? Why does Joseph give up trying to be her friend? Why do you think lots of kids ignore students like Shirley, rather than try to befriend them?
- ✦ What is it about Shirley that makes Mabel decide to be her friend? What does Mabel teach Shirley that helps her to become accepted by the other kids?

- ✦ Can you think of other "heroes" who, like Jackie Robinson, became successful, overcoming obstacles of racial or cultural prejudice? The person does not have to be famous—he or she could be someone in your family, school, or neighborhood. What character traits helped this person to be a success?
- ✦ Discuss the meaning of the word *diversity.* Is your school diverse? Would you rather go to a school where all the students have similar backgrounds, or one with a more diverse group of students? What might be the advantages of going to school with kids from many different cultures?
- ✦ Imagine that your family moves to a foreign country with customs quite different from those you are used to. What do you think you would miss most about your present life?
- ✦ Think about the story Shirley tells about the fisherman's wife and the filial (faithful) daughter. Why does the story have special meaning for Shirley? Why would it be meaningful for any person who left her or his native land to make a new life in a new country? Have you had an experience that makes the story especially meaningful to you?

ABOUT THE AUTHOR: Betty Bao Lord was born in Shanghai, China, in 1938, and came to America when she was eight years old. Her immigrant experiences heavily influence *In the Year of the Boar and Jackie Robinson.* She married an American diplomat, Winston Lord, enabling her to return to China in an official capacity. Her visits to China were the inspiration for many of her other books, including two adult novels: *Legacies: A Chinese Mosaic* and *The Spring Moon: A Novel of China.*

Beyond the Book...

CHINESE NEW YEAR: Find out more about Chinese New Year and other traditional Chinese holidays, such as the Mid-Autumn Festival that Shirley celebrates with her parents. Celebrate one or more of these holidays in the traditional way. Talk about what you can learn about Chinese culture through learning holiday traditions. For example, what values are more important in Chinese culture than in U.S. culture?

BILINGUAL EDUCATION: Go to the foreign language or English as a Second Language department in your school and ask if they have any information on bilingual education. Talk about what it would be like to grow up speaking two different languages (or what it has been like, if you come from a bilingual family).

GUEST SPEAKER: Invite an adult or young person who has come to the United States from another country to speak to your group. Ask your speaker to talk about differences between her or his culture and U.S. culture, and any difficulties she or he may have had adjusting. Schedule a question-and-answer period after the talk.

FOLKTALES: Find and collect folktales from your own heritage that relate to your life, as the story of the fisherman's wife and the filial daughter relate to Shirley's. Tell your story and explain why you find it meaningful.

JACKIE ROBINSON: Find out as much as you can about Jackie Robinson. Biographies of Robinson will probably be available in your library. Discuss the obstacles he had to overcome and why he is considered a hero.

BASEBALL: As a family, go to a baseball game. Do some research on baseball to discover why it has come to represent American values.

CHINESE CHARACTERS: The Chinese characters for the months are used for the titles of the chapters in this book. Try writing these characters. If you know someone who knows Chinese, ask them to show you how, or look it up in a book. Make a monthly calendar using what you've learned.

REFRESHMENTS OR FOOD MENTIONED IN THE BOOK: Serve candied plums or orange noodles out of a can. If you can find them, serve Chinese moon cakes filled with lotus seeds and honey.

IF YOU LIKED THIS BOOK, TRY...

Elaine, Mary Lewis, and the Frogs, by Heidi Chang—A Chinese-American girl who moves to Iowa has difficulty adjusting until she meets a girl who loves frogs.

Felita, by Nicholasa Mohr—A Puerto Rican family moves to an area where Spanish is not spoken.

Some Other Books by Bette Bao Lord:

The Spring Moon: A Novel of China

The Middle Heart: A Novel

Island of the Blue Dolphins

by Scott O'Dell

When Karana, a young Indian girl, is stranded by her people and forced to live by herself on a deserted island, she discovers how to draw upon her skills and strengths to survive.

I'm not sure I would have survived alone for as long as Karana did. Girls can learn a real lesson about being alone and finding strength within—things the distractions of TV, radio, and computers tend to take away.

READING TIME: 2–3 hours, about 181 pages
THEMES: loneliness, gender roles, love, grief, independence

Book Discussion Questions

- Why did Ramo feel that his father's death meant that he was chief of the tribe? How did Karana feel about him taking on her father's role? Did she think he was ready? Do you? Why or why not?
- How does Karana deal with her brother's death? Why does she decide to leave her village and move farther into the wild? What do you think you would have done?
- Why does Karana burn the village after Ramo dies? Are there any places that trigger painful memories or associations for you? Why? How do you handle these feelings?
- Women in Karana's tribe were forbidden to make weapons and her father and other members of the tribe told her it was wrong. Why does she now persevere in that task? What does that tell you about her character? Do the bad things she was warned of actually ever happen?
- What sustains Karana in her isolation? Imagine being in her situation. What would you have drawn upon to exist in such solitude?
- Karana, on her ill-fated attempt to search for her people, says, "I was not nearly so skilled with a canoe as these men, but I must say

that whatever might befall me on the endless waters did not trouble me. It meant far less than the thought of staying on the island alone, without a home or companions, pursued by wild dogs, where everything reminded me of those who were dead and those who had gone away." When does Karana realize that this is not true?

- ✦ Why is the appearance of the dolphins important to Karana?
- ✦ What made Karana decide that she would not leave the island until the white men came back for her?
- ✦ Why does Karana care for the dog she had tried to kill? How does nursing him back to health change her feelings towards him?
- ✦ How does trapping and killing the devilfish change Karana?
- ✦ The Aleuts, her enemies, return to the island after several years. Karana observes, "I had not heard words spoken for so long that they sounded strange to me, yet they were good to hear, even though it was an enemy who spoke them." What do you think she means by this? Why does Karana consider the girl an enemy even though she is friendly to Karana?
- ✦ How does her friendship with the Aleut girl, Tutok, affect Karana? How is her feeling about living alone on the island affected by having had a friend?
- ✦ What does Karana learn by befriending the animals on the island? What do those friendships offer her?
- ✦ What's surprising about Karana's response to the wild dog Rontu's death?
- ✦ How does Karana change during the course of the book?
- ✦ What do you think would have happened to Karana when she rejoined people? What kind of adjustment would she have had to make?
- ✦ How many of the beliefs that Karana held at the beginning of the book turned out to be false by the end?

ABOUT THE AUTHOR: Scott O'Dell was born in 1898 in Los Angeles and did many things in his 91 years. Besides attending Occidental College, University of Wisconsin-Madison, Stanford

University, and the University of Rome, he worked in film production for several years as a cameraman and a technical director, and also worked as the book editor for a Los Angeles newspaper. He served in the U.S. Air Force during World War II, then went on to write numerous, well-loved books for children and young adults. He received several awards for his writing, and established the Scott O'Dell Award for Historical Fiction. Mr. O'Dell died in 1989.

Beyond the Book...

SEASHORE: Visit a beach or seashore, if there is one nearby, and gather seashells. Go at a time when the beach is usually empty and sit and listen to the sounds of the ocean. Try to imagine what it would be like to be alone there for an extended period of time.

DOLPHINS: Do some research on how dolphins communicate and their natural habitat. Then go to a local zoo or aquarium and see how the dolphins live there. While you're there, speak with the zookeeper and ask him to tell you about the individual dolphins who live there.

MAP: Based on the way the island is described in the book, create a map. Color it in and mark the places that were important to Karana as she made a life on the island.

JEWELRY-MAKING: Gather some seashells from a local beach or some rocks or beads from a craft or bead store. Make necklaces from these beads and rocks.

REFRESHMENTS OR FOOD MENTIONED IN THE BOOK: Prepare scallops or other shellfish. Try some seaweed (the kind you can get in Asian restaurants or markets).

IF YOU LIKED THIS BOOK, TRY...

Hatchet; Dogsong, by Gary Paulsen—These are other "survival" stories (see p. 67).

Some Other Books by Scott O'Dell:

Sarah Bishop

My Name is Not Angelic

Jacob Have I Loved

by Katherine Paterson

In this coming-of-age story, an outspoken, spirited, and sometimes prickly island girl longs for an identity apart from her pampered and pleasing twin sister. The wild environment of the Chesapeake Bay offers the heroine the challenge she seeks to become her own particular person and find a purpose in life beyond the expectations of her family. This beautifully written book raises important issues about sibling relationships.

READING TIME: 3–4 hours, about 175 pages
THEMES: coming of age, work, sibling rivalry, gender roles, identity

Discussion Questions

- Louise's difficult grandmother spits out a biblical quotation that nearly tears the young girl apart: "Jacob have I loved, but Esau I have hated. . . ." Why is this quotation so painful to Louise?
- Compare the temperaments and personalities of Caroline and Louise Bradshaw.
- Louise tells her side of her family story in an impatient, resentful voice. Describe some of the past family events and memories that trigger off Louise's feelings of resentment. What seems to ail her so much?
- How does Louise feel about her nickname "Wheeze?" If you have a nickname, how do you feel about it?
- Louise loves the water that surrounds Rass Island, though "...the women of my island were not supposed to love the water. Water was the wild, untamed kingdom of our men." What does the water of her beloved Chesapeake Bay signify to Louise?
- Describe Louise's youthful relationship with her friend Call. What is their relationship as they approach adulthood?

- Louise confesses to an abiding hatred of her sister. Do you think brothers and sisters sometimes feel such deep resentment for each other? Do you think this is normal?
- Louise's powerful feelings toward her sister are bottled up in her heart. Should she have taken more responsibility for improving her relationship with Caroline? In what specific ways might she have done this?
- Describe Louise's feelings for The Captain.
- What are some of young Louise's conflicted feelings about being female? How does she come to terms with her femininity after she becomes an adult?
- Louise talks about one winter that gave her the happiest days of her life. Share some special memories of the happiest days of your own life.
- Louise's mother confesses that: "...I am what I wanted to be. ...No one made me become what I am." How are Louise's life choices similar to those her mother made?
- Louise says: "...I decided that if you can't catch crabs where you are, you move your pots." What does this mean?
- What do you think of the choices Louise makes in her adult life?
- Do you believe that Louise has made peace with her twin sister, Caroline, by the time she reaches adulthood?

ABOUT THE AUTHOR: When Katherine Paterson was growing up she wanted to be a movie star. Or a missionary. Or a mother. She never wanted to be a writer—until she was one. Her literary career began with the Presbyterian church in 1964, when she was hired to write some curriculum materials for fourth-, fifth-, and sixth-graders. Out of this work grew a passion for writing that Ms. Paterson couldn't ignore. But she had a lot to learn. Over the course of four years, she wrote and wrote and wrote (between feeding, diapering, and carpooling her four children) but had no success. So she enrolled in a creative writing course, and the novel she wrote for the class was eventually published. Since then, she has created numerous children's and young-adult novels. When she's not writing, Ms. Paterson loves to sing and play the piano. She also loves to spend time with her two granddaughters.

Beyond the Book...

SPECIAL PLACES: The special Chesapeake Bay location is as much a part of Louise as the color of her eyes. If you feel a strong bond to a particular place in your memory—a summer vacation spot, a long-ago childhood home—find some pictures of that place to share with each other. Discuss why certain places become part of our memories. Is it the place itself? The experiences we had in that place? The people who were there? How do certain places become part of who we are?

CHESAPEAKE BAY: The word Chesapeake is an Algonquin Indian name that means "great shellfish bay." Find a seafood chowder recipe in a cookbook. Make the soup and serve it to a friend for dinner.

TWINS: Invite a set of twins to talk about their experience of twinship. Ask whether anyone knows twins or went through a stage of wishing they had a twin. Encourage speculation on what it must be like to carve out an individual identity if one is a twin, or how one develops as an individual within peer groups who enjoy doing the same things—dressing similarly, sharing the same tastes.

IDENTITY BOXES: Take plain cigar boxes and decorate them inside and out with items that represent you. For example, if you are a dancer, use a doll's ballet slipper, or if you love the beach, use a seashell. By the time you're finished, your best friend should be able to pick out your box.

REFRESHMENTS OR FOOD MENTIONED IN THE BOOK: Find a recipe and try preparing crab soup. Learn to shell and make fresh peas.

IF YOU LIKED THIS BOOK, TRY...

Beautiful Swimmers: Watermen, Crabs, and the Chesapeake Bay, by William W. Warner—Katherine Paterson said that this Pulitzer Prize–winning book, written in 1976, was the inspiration for the Chesapeake Bay setting of *Jacob Have I Loved.*

Bridge to Terabithia, by Katherine Paterson—This novel also features themes of loss, identity, heartbreak, and growing up (see p. 21).

Come Sing, Jimmy Jo, by Katherine Paterson—A young boy must figure out how he fits into his family.

The Jellyfish Season, by Mary D. Hahn—A young girl copes with family problems and change when she moves to the Chesapeake Bay.

Some Other Books by Katherine Paterson:

The Great Gilly Hopkins (see p. 113)

Jip: His Story

Lyddie

Julie of the Wolves

by Jean Craighead George

This novel of courage and survival tells the story of 13-year-old Julie, an Eskimo girl, lost in the Alaskan wilderness. Out on the pitiless tundra, Julie draws on everything she has ever learned from her Eskimo culture in order to communicate with a pack of wolves who, she knows, will protect her from starvation and cold. While her knowledge of Eskimo ways eventually saves her in the wild, Julie must decide whether to rejoin the more "civilized" world of modern assimilated Eskimos or return to nature and follow the ways of her ancestors.

READING TIME: 3–4 hours, about 170 pages
THEMES: survival, father-daughter relationships, nature, cultural identity, self-discovery

Discussion Questions

✦ How does Julie's absent father help her to survive on the tundra?

✦ Discuss the different ways Julie makes herself wolf-like.

✦ Julie's father has taught her to understand animal gestures in order to communicate with them. Discuss how your own pets express goodwill and friendship. How do they express aggression? Submission?

✦ Julie's father told her that animals have languages of their own to express themselves. What sounds do animals make to communicate fear, warning, affection, nervousness?

✦ How does Julie feel about the marriage that has been arranged for her? Specifically, what aspects of the marriage was Julie running from? What do you think of the custom of arranged marriages?

✦ In traditional Eskimo culture, the children learn the values and ways of their ancestors to help them survive their harsh climate.

What skills and values from your parents and grandparents help you survive in the world you live in?

- Hundreds of interesting facts about Eskimo life and wolf behavior are woven into the story. What memorable facts did you learn from the book?

- Julie tells one of the wolves: "We Eskimos have joking partners—people to have fun with—and serious partners—people to work and think with." Talk about friendships you have that are based mainly on fun and other relationships that are more serious.

- How did you react to some of the hunting scenes? In what ways does Julie show respect for the animals she must eat in order to survive?

- How does Julie feel about losing the nickname Meets and becoming Julie? What does her "civilized" name represent for her?

Reading this thrilling adventure story gave me a new respect for nature—and for the wolf.

- Julie remembers her father telling her of the importance of wolves to the great chain of life. Discuss how the other animals like the caribou, lemmings, birds, and weasels depend on the wolves for their own survival.

- What were your feelings about the senseless death of the great wolf, Amaroq?

- The wisdom of her absent father saves Julie's life. Why does she choose to leave him in the end?

ABOUT THE AUTHOR: Jean Craighead George was born in Washington, DC. She graduated from Pennsylvania State University with degrees in both English and science. The time she has spent observing and reporting on animal behavior in the wild has enabled her to write numerous nature books for children, including the Newbery Medal–winning *Julie of the Wolves.* Many of the characters she creates are based on the numerous pets and wild animals she lives with in her home in Chappaqua, NY. Ms. George is a mother of three, and a doting grandmother who enjoys reading to her grandchildren every chance she gets. She enjoys hiking, canoeing, and making sourdough pancakes.

Beyond the Book...

SURVIVAL: Divide into teams. Give each team a list of 20 items one would need living alone in the wilderness, and then tell them they can have only eight. Have each team work together to decide which are the most important. Once everyone has completed the task, talk about how those decisions were made.

MAP: Make available a map of Alaska and find the North Slope location where Julie was lost. Try to imagine—really imagine—what this experience must have been like. What's the coldest you've ever felt? Hungriest? Thirstiest? What was your most difficult physical ordeal?

MOVIE: See the film *Never Cry Wolf,* about a scientist sent to live on his own in the Arctic to study the wolves in that region.

REFRESHMENTS OR FOOD MENTIONED IN THE BOOK: Some of the foods mentioned are stew, wild plants, and dried meat. Research recipes that use edible wild plants and prepare a dish made from them to share. (Note: Eating or using wild plants can be dangerous. Consult a naturalist or field guide if you are foraging for wild edibles.)

IF YOU LIKED THIS BOOK, TRY...

My Side of the Mountain, by Jean Craighead George—An adventure story about a young boy who passes a winter in the Catskill Mountains by himself.

Pack, Band, and Colony: The World of Social Animals, by Judith and Herbert Kohl—A nonfiction look at animals that live in groups.

Wolf Pack: Tracking Wolves in the Wild, by Alice Aamodt and Sylvia A. Johnson—This book examines the social structure and habits of a wolf pack.

Some Other Books by Jean Craighead George:

The Missing Gator of Gumbo Limbo

The Fire Bug Connection

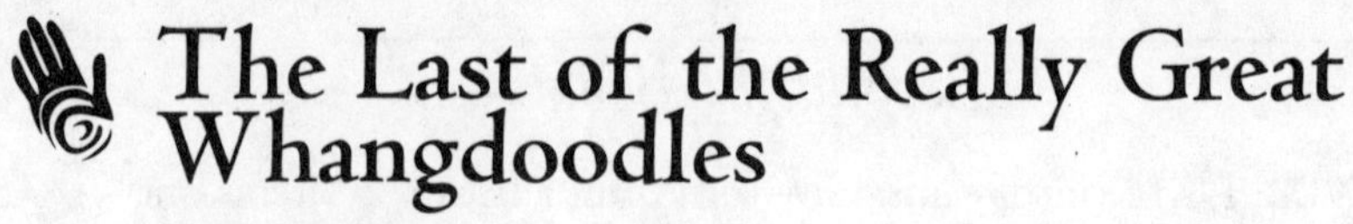

The Last of the Really Great Whangdoodles

by Julie Edwards

Ben, Tom, and Lindy Potter meet the eccentric Professor Savant, who tells them about the Whangdoodles—magical creatures who once lived on earth but then disappeared to another land. The Professor promises the children that, if they learn to *really* use their imaginations and their five senses, they will be able to visit Whangdoodleland and meet the great Whangdoodle himself. After braving dangers of the most fantastic sort, Ben, Tom, and Lindy achieve their goal.

READING TIME: 2–3 hours, about 275 pages
THEMES: imagination, creativity, courage, faith, trust, fantasy

Discussion Questions

- Professor Savant teaches Ben, Tom, and Lindy how to appreciate the beauties of the world that most people miss, because "Nobody thinks to look." How is this true in the book? What are some of the beautiful things in your own world that you can see only if you *really* look?

- Think about all the times Lindy is afraid, but overcomes her fears. What would she have missed if she had let her fears control her actions? Have you ever done something that you were afraid to do? How did you feel afterwards? What might you have missed if you had let your fear stop you? Talk about how you managed to overcome your particular fear.

- What lesson do the children learn from the Sidewinders? What are the "Sidewinders" in your life? Can you remember a time when you remained calm and steady, even though you were nervous or worried about something? Share your experience.

- What lesson do the boys learn from the Gazooks? What are the "Gazooks" in our world? Can you name things that commercials and other advertisements make us want, even though they may not be good for us?

> *I love to travel, and though I have been to some really wonderful places and had some amazing experiences, none of them compares to the incredible trips imagined in this book!*

- When the Swamp Gaboons make fun of Tom, the Professor says, "I warned you not to answer back. It only encourages them." Do you know any "Swamp Gaboons"—rude people who like to tease others, or make fun of them? If you try to answer a Swamp Gaboon back, what usually happens? What is the best way to deal with people who tease you or make fun of you?
- Who is your favorite character? Why?
- Prock does everything he can to keep the Professor and the children from meeting the Whangdoodle. Are there any "Procks" in your life—things that make you think you can't achieve your goals? Tell about a time you overcame difficulties to achieve a goal. How did you do it?
- Even though the Professor can't see the bridge to the Whangdoodle's palace, Lindy helps him across by telling him to not look down, and describing the wonderful palace on the other side. What lesson can readers learn from this? Is there a difficult goal you would like to achieve? How can negative thoughts keep you from achieving it? How can concentrating on your goal help you to achieve it?

ABOUT THE AUTHOR: Julie Andrews Edwards is widely known as the star of such classic films as *Mary Poppins, The Sound of Music, Thoroughly Modern Millie,* and others. She also starred in the Broadway musicals *My Fair Lady, Camelot,* and *Victor/Victoria.* Besides her highly acclaimed career in show business, Ms. Edwards is on the board of Operation USA, an international relief organization, and the Foundation for Hereditary Disease, with her husband Blake Edwards.

Beyond the Book...

USING YOUR SENSES: Practice experiencing the world as if you had studied with Professor Savant. Make a "Five Senses Chart" with five columns headed "See," "Hear," "Touch," "Taste," and "Smell." Pick a place to explore, or just sit and observe. It could be anywhere outdoors, or even in your own room. Make a special effort to notice everything around you, using all your five senses. Write down the things you see, hear, feel, taste, and smell.

ART: Use the vivid descriptions of Whangdoodleland and the fantastic creatures the children meet there to draw or paint a picture. You can stick to the book, or, if you prefer, draw your own vision of Whangdoodleland or some other fantastic place you make up yourself.

ALLEGORY: The Last of the Really Great Whangdoodles is an allegory, or story that teaches a lesson—in this case, that your imagination can fill your world with magic and wonder. Like most allegories, the characters have adventures that help them learn the main lesson of the story. Try writing your own allegory. You can use Professor Savant's lesson, or any other you would like your story to teach. (Hint: You might want to include a character who, like the Professor, helps your other characters—and your readers—learn your story's lesson.)

IF YOU LIKED THIS BOOK, TRY...

The Chronicles of Narnia, a series by C. S. Lewis—In these six books, children travel to a magical land, have fantastic adventures, and learn important lessons about life. The books are: *The Lion, the Witch, and the Wardrobe* (see p. 166); *Prince Caspian; The Voyage of the "Dawn Treader"; The Silver Chair; The Horse and His Boy; The Last Battle; The Magician's Nephew.*

Chronicles of Prydain, by Lloyd Alexander—In this five book series, Young Taran saves his beloved land from the forces of evil. The books are: *The Book of Three, The Black Cauldron, The Castle of Lyr, Taran Wanderer, The High King.*

Some Other Books by Julie Edwards:

Mandy

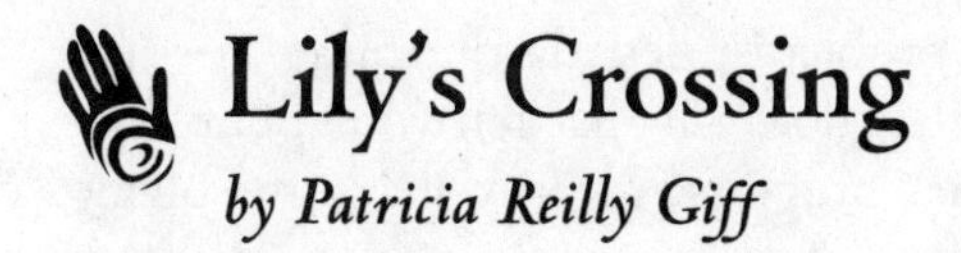

Lily's Crossing

by Patricia Reilly Giff

It is the beginning of the summer of 1944. Lily Mollahan has nothing more strenuous planned than going to the family beach cottage with her father and grandmother as she has every summer. But this is 1944, and World War II is going on overseas. The war takes Lily's beloved father away as well as her best friend. She is left with Albert, a lonely Hungarian refugee boy who has lost most of his family in Europe. Lily grows up fast in unexpected ways during this wartime summer in Rockaway, NY.

READING TIME: 2–3 hours, about 181 pages
THEMES: loss, separation, war, friendship, coming of age

Discussion Questions

- ✦ Lily describes the delicious feeling of a carefree summer stretching ahead—no homework, no piano practice—just writing stories, sneaking into the movies, and going fishing. Share some summer memories of specific things you like to do—or stop doing—when summer comes.
- ✦ Have you ever had a summer that turned out differently than you planned?
- ✦ How are Lily's feelings about school similar to or different from some of yours?
- ✦ What memories does Lily have of her mother?
- ✦ How does Lily get along with her grandmother?
- ✦ Like many grown-ups, Lily's grandmother tells her: "You can do anything if you really work at it." Discuss what you think about this advice that adults often give to young people.
- ✦ Although the United States fought World War II overseas, the war affected daily life at home in big and small ways. In what ways does the war intrude on Lily's carefree summer life?

- ✦ Lily makes a list of her life problems: lies, daydreaming, needing friends, her grandmother. What does this list tell you about Lily? By the end of the book, what progress has Lily made in solving the problems on her list?
- ✦ Lily likes stretching the truth—sometimes a lot. What are some of the fibs, white lies, and outright whoppers she tells? Why do you think kids sometimes get into the habit of lying?
- ✦ Lily's neighbor Mrs. Orban introduces her nephew Albert in this way: "Isn't this perfect. ...Just as Margaret leaves, Albert comes." How does Lily feel when her best friend, Margaret, says she's moving away? What does she think of Albert as a substitute for Margaret?

Patricia Reilly Giff's storytelling is so vivid, I felt I was actually in the book.

- ✦ Why do you think the ocean and the water are so important in this book?
- ✦ How have the summer's unexpected turns changed Lily by the end of the book?

ABOUT THE AUTHOR: Patricia Reilly Giff has been surrounded by books from the moment she first picked one up. She has written over 60 books, taught reading and writing, and opened a bookstore. Giff says that her involvement with teaching stems from her desire to share her love of reading with others. This is also the reason she chose to write—she wants to give her readers a reason to get lost in a book and to experience the pleasure that reading can provide. Many of her books are based on her own experiences, or those of children she has known. Her own children, Ali, Jim, and Bill, all now adults, often appear in her books. Giff lives with her husband and three cats in Weston, CT. She spends her time lecturing, working at the family-owned bookstore, going to the movies (where she loves to eat popcorn), and continuing her quest to bring books and children together.

Beyond the Book...

INTERVIEWS: World War II touched the lives of many Americans in large and small ways. To understand what wartime America felt like during the last few years of the war, interview relatives, neighbors, or friends who were alive in 1944. Ask them, in particular, to recall what daily life was like for young people at that time and the different ways the war overseas affected them. Share these recollections with each other. If any of the subjects have wartime letters to share, read them aloud.

MOVIES: Rent the Academy Award–winning movie *The Best Years of Our Lives,* a bittersweet film about the return of American soldiers to their families. The movie takes place during the same time period as *Lily's Crossing.* Afterwards, discuss similarities between the movie and the book.

REFRESHMENTS OR FOOD MENTIONED IN THE BOOK: Make a dinner of flounder, corn on the cob, and lemon cake. Or you may want to gather all the candy they talk about in the book: Milky Ways™, Lifesavers™, Necco™ Wafers, and Hershey™ Bars. Share a bowl of penny candies mentioned in the book, many of which are available today—though not at penny prices!

IF YOU LIKED THIS BOOK, TRY...

Salted Lemons, by Doris Buchanan Smith—This story, about a girl who moves to a new town is set during World War II.

The Silver Sword, by Ian Serraillier—During World War II, two Polish children are reunited with their parents after a painful separation.

Summer of My German Soldier, by Bette Greene (see p. 244).

Some Other Books by Patricia Reilly Giff:

Have You Seen Hyacinth Magaw
Left-Handed Short Stop
Dance with Rosie

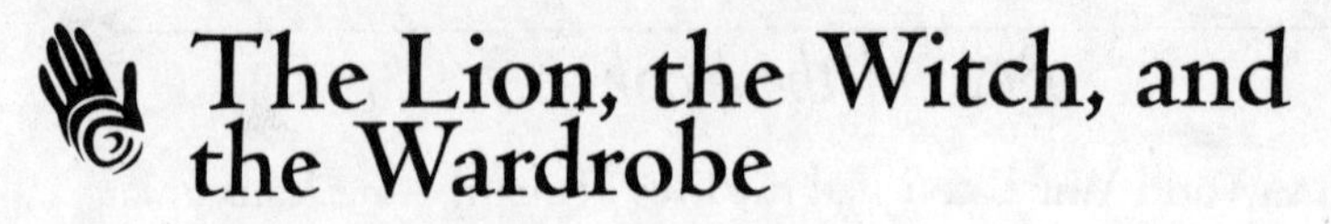

The Lion, the Witch, and the Wardrobe

by C. S. Lewis

In this powerful fantasy of the triumph of good over evil, four English children enter the enchanted land of Narnia through a wardrobe. Leaving ordinary time and their everyday childhood world behind, the four children confront and vanquish the evil White Witch of Narnia to save its inhabitants. In doing so, they achieve full heroic adult stature in Narnia before returning, through the wardrobe once more, to their childhood.

READING TIME: 3–4 hours, about 186 pages
THEMES: good versus evil, faith, compassion, courage, coming of age, rebirth

Discussion Questions

- In his dedication to his godchild, C. S. Lewis says: "But some day you will be old enough to start reading fairy tales again." What ages do you think are the most enjoyable for reading fairy tales? Have you outgrown them? If so, why? If not, what makes you continue to enjoy fairy tales?
- The four children, Peter, Susan, Edmund, and Lucy, are introduced without any last names. Their parents are never mentioned. Their former home is never described. Why do you think the author provides so few background details about the children?
- At first the children are restless. The always impatient Edmund says: "Of course it would be raining!" Why do you think it's important that this tale begin on a rainy day? Discuss any imaginary games you may have played on long, rainy days.
- When they first meet Aslan, the children are struck by his appearance: "People who have not been to Narnia sometimes think that a thing cannot be good and terrible at the same time. If the children had ever thought so, they were cured of it now." What about

Aslan was both "good and terrible"? Have you ever encountered something you felt that way about?

- ✦ When you were younger—or perhaps even now—did you ever have a secret place indoors or outdoors where you played fantasy games? If so, describe the place and the qualities that made it a special hideaway.
- ✦ Share with everyone any fantasy worlds that you created during your childhood. Was there a point when you gave up playing imaginary games? If so, do you remember how and why you gave them up? Why do you think fantasy play is so important to children?
- ✦ What are Lucy's feelings when she sees fir trees instead of fur coats halfway between the wardrobe and Narnia?
- ✦ Share your impressions of each of the four children—Peter, Susan, Edmund, and Lucy.
- ✦ What is it about Edmund's character that attracts him to the White Witch?
- ✦ Can you think of any other fantasies or fairy tales in which a character makes a terrible mistake, as Edmund does, and takes some kind of magical food or potion?
- ✦ Why do you think Edmund is in the story at all? In what ways do you think he is necessary to the story?
- ✦ The narrator, who often addresses the reader, says: "Perhaps it has sometimes happened to you in a dream that someone says something which you don't understand, but in the dream it feels as if it had some enormous meaning—either a terrifying one which turns the whole dream into a nightmare or else a lovely meaning too lovely to put into words, which makes the dream so beautiful that you remember it all your life and are always wishing you could get into that dream again." If you have ever had this kind of powerful dream, and wish to share it, please tell it.
- ✦ Figures similar to the White Witch have appeared in many fairy tales, fantasies, and movies. Compare the White Witch of Narnia with some of the witches you love to hate in other books or in movies.
- ✦ The children are always on the move in Narnia. Compare the forest

setting they first enter to their final destination of Cair Paravel. What feelings do you associate with each setting?

- ✦ What do you think of the idea that many years pass in Narnia, and the children grow completely into adults, while on the "real world" side of the wardrobe, not a minute has passed?
- ✦ If you believe in an afterlife, what parts of Narnia are similar to what you imagine an afterlife to be?
- ✦ Aslan, the golden lion and force of goodness in Narnia, seems to save the children. Yet in the end, the children save Aslan. What do you think the author is trying to say by having the children save the savior in the story?
- ✦ Imagine that a door in the room in which you are sitting right now leads to a fantasy world. What would that be like? Share your fantasy.

ABOUT THE AUTHOR: Clive Staples (C. S.) Lewis (often known as Jack) was born in Belfast, Ireland, on November 29, 1898, and was brought up in a home filled with books. Mr. Lewis and his brother Warren were avid readers who sometimes felt books were more real than the actual world. C. S. Lewis was ten when his mother passed away and he sought refuge in his writing. In his lifetime, Mr. Lewis wrote more than 30 books, from children's literature and science fiction to religion.

Beyond the Book...

SETTING: The Lion, the Witch, and the Wardrobe is filled with descriptions of its extraordinary settings. Prior to the discussion of this book, decide on a favorite scene to paint or draw. While working on the project, talk about the significance of the various seasons and settings in the book. Speculate on why certain characters are associated with one setting or climate rather than another.

MOVIE: Watch the Public Broadcasting System film *The Chronicles of Narnia.* Discuss whether the characters and setting do justice to the book or whether you envisioned things differently. Consider if movie versions of favorite books enhance or detract from your reading experience.

REFRESHMENTS OR FOOD MENTIONED IN THE BOOK: Serve a tea with tea sandwiches of butter and ham, trout, buttered potatoes, and Turkish Delight candy. During the tea, think about what other stories feature characters who eat magical foods.

IF YOU LIKED THIS BOOK, TRY...

The Chronicles of Narnia, a series by C. S. Lewis—The characters from *The Lion, the Witch, and the Wardrobe* journey into Narnia again for further adventures.

The Book of Three, by Lloyd Alexander—A young farmer becomes a hero on his quest through a magical land (winner of 1969 Newbery Medal).

Little House on the Prairie

by Laura Ingalls Wilder

This volume in the eight-book *Little House* series tells the story of the covered-wagon journey the Ingalls family makes as they travel from their little cabin in Wisconsin to the Kansas plains, where they build a new home. Life on the prairie is full of new experiences like encounters with wolves, contact with Indians (both friendly and tense), a prairie fire, and family sickness. The Ingalls family survives them all. What they cannot survive, however, is a government order that all settlers must move, due to increasingly tense encounters with the Indians. The Ingalls family packs up their covered wagon again to head even farther west.

READING TIME: 4–5 hours, about 340 pages
THEMES: growing up, moving, resourcefulness, separation, hardship, family

Discussion Questions

- Laura and Mary Ingalls never complain about moving from their snug Wisconsin cabin. They are described simply as clinging tightly to their rag dolls. How do you think they feel about leaving their familiar home, friends, and relatives and heading into the unknown?

- Describe your feelings and thoughts about moves you have made in your own life from one home to another. What were the worst and best parts of the experience? What surprised you about the new setting? What helped you to adjust?

- Describe the behavior of the Ingalls children on their difficult journey. How does their journey compare to a long car, plane, or train trip a child of today would make?

- What possessions did the Ingalls family bring along on their trip? How are they different from what people bring when they move today?

- ✦ What personal and practical skills does each family member need on a covered-wagon trip?
- ✦ Each family member, including the family animals, has a role to play in the Ingalls' household. Describe what those roles are. Why do you think such roles are established? Are there distinct roles in your own family?
- ✦ Laura Ingalls is deeply moved by the physical beauty of the prairie. Share your impressions of the book's prairie setting. Does it sound like somewhere you would like to live?
- ✦ What are some of the simple activities the Ingalls family enjoys together? Describe activities you enjoy with your own family.
- ✦ Describe the relationship between the Ingalls family and the neighboring Indians.
- ✦ How do each of the Ingalls family members react to the news that they must leave their new prairie home?
- ✦ What is the mood of the family as they begin their second journey toward an unknown land?

ABOUT THE AUTHOR: Laura Ingalls Wilder's life was much like that of the girl she depicts in her books. Ms. Wilder was raised in the late 1800s, and her family faced many hardships and was forced to move many times throughout the Midwest frontier. Ms. Wilder drew from these experiences to create a series of books, for young adults, about her life. These books chronicle the struggles her family encountered as they fought to survive in a new land. Her memories make this series of books come to life. Ms. Wilder became a teacher and she and her husband, Almanzo, and their daughter, Rose, settled in Missouri. Rose became a newspaper writer and helped her mother publish her first book, *Little House in the Big Woods.* Seven books followed. Ms. Wilder died in 1957 at the age of 90.

Beyond the Book...

MAP: Before the book discussion, look at a map of the United States and trace the journey of the Ingalls family from Wisconsin, through Minnesota, Iowa, Missouri, and Kansas. Imagine yourselves on the trip and write one or two diary entries from the point of view of Laura or her mother. What work was involved? How did the pioneers deal with bathing, going to the bathroom, cleaning clothes? Share your entries during the book discussion.

HISTORICAL SOCIETY: You may enjoy a visit to the local historical society to view documents, photos, clothing, and objects from the same time period as that of the *Little House* books. Share your own period documents, like old family photos, antique objects, and old letters, as a way of re-creating an earlier time. Try to create your own "antique" documents: write diary pages or letters as if you were living during the 1870s, then soak them in tea.

PEPPERMINT CANDIES: Have a bowl of peppermint candies on hand to pass around at your book discussion. Talk about how this simple, everyday treat, which can be bought just about anywhere, was an amazing and delicious surprise for the Ingalls girls because of its rarity.

REFRESHMENTS OR FOOD MENTIONED IN THE BOOK: If your home has a working fireplace, try cooking a meal (such as a stew) in a thick iron skillet or pot over a roaring fire.

IF YOU LIKED THIS BOOK, TRY...

Little House on the Prairie series, by Laura Ingalls Wilder—These stories continue the tale of the Ingalls' family life on the frontier.

One Day on the Prairie, by Jean Craighead George—This nonfiction book is about a day the author spent at the Prairie Wildlife Refuge in Oklahoma.

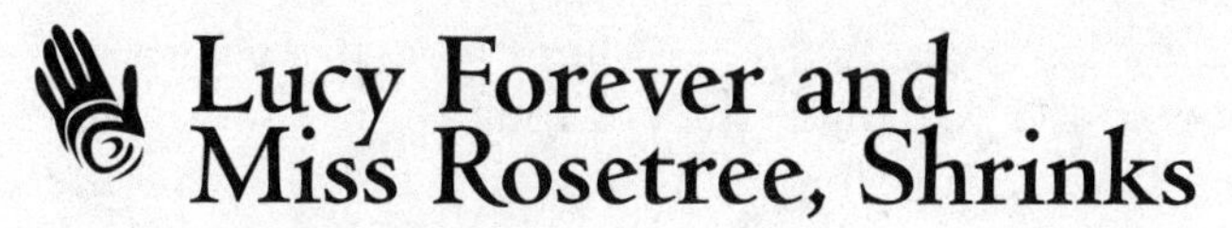

Lucy Forever and Miss Rosetree, Shrinks

by Susan Shreve

Lucy and Rosie, two sixth graders with a pretend psychiatry practice, love to invent ridiculous psychiatric case histories. But things get more serious when they come upon a small, mute girl from an orphanage and determine they are going to use all their knowledge to help her to talk. In doing so, they discover that little Cinder is an abused child, and rescue her from her abusive caretakers.

I met author Susan Shreve last summer on Martha's Vineyard at a dinner hosted by her brother. We spent a wonderful evening together, and I left excited and eager to read her books. This one raises some tough issues—child abuse, troublesome family relationships—but it does so in an accessible, insightful way.

READING TIME: 1–2 hours, about 120 pages
THEMES: child abuse, loneliness, siblings, parent-child relationships, honesty, friendship

Discussion Questions

✦ How do Lucy and Rosie figure out what's happening with Cinder? Have you heard or read anything about child abuse? What do you think qualifies a parent's behavior as abusive? Do you think a child can be abused by words only, without being physically hurt? What would you do if you thought you knew that a child was being abused?

✦ Lucy says that she feels rejected by her parents because her mother and father are so close that they often make her feel left out. From what you read, do you think Lucy's parents really love her? If Dr. and Mrs. Childs asked you for advice, would you have ideas on how they could be better parents? What would you say to them?

- ✦ Lucy knows that lying is wrong, yet she lies to her parents several times in this book. Do you think she is wrong to lie about going into her father's files, quitting piano lessons, or going to the orphanage? Do you agree that "honesty is the best policy" all the time? Do you think it's okay to lie sometimes? When are those times? What makes those times different?
- ✦ Mrs. Treeman keeps close tabs on Rosie, and thinks that the Childs give Lucy "too much freedom." Do you agree? Which method of child-raising do you think is preferable—the Treeman's, the Childs's, or something in between?
- ✦ Daughters: Are you an only child? Do you ever share Lucy's feelings of loneliness and wish for a sister or brother? Or, do you have many sisters and brothers? Do you ever share Rosie's wish for more privacy and time to be alone? Mothers: If you feel comfortable discussing this matter, talk about your choice to have only one child, or to have a large family.
- ✦ What do you think about the way Lucy talks to and about Mr. Van Dyke? For example, when he threatens to call her mother, Lucy says, "You can call her every hour if you'd like." Why does she react to him this way? Do you think it's okay for kids to speak to adults this way, or do you think young people should be more respectful, even when the adult is being rude or mean? Is there anyone you've wanted to stand up to the way Lucy stands up to Mr. Van Dyke? Did you do it?

ABOUT THE AUTHOR: Susan Shreve was born in 1939 in Toledo, OH. Like Robert Louis Stevenson, much of her young childhood was spent in bed, fighting illness and boredom. To entertain herself during her bout with polio, she invented various scenarios for her dolls, which have made their way into her books in various forms. Ms. Shreve overcame the hardships of her young life. She graduated from the University of Pennsylvania magna cum laude, married, had four children, became both a teacher and a writer, and is recognized as a major American novelist.

Beyond the Book...

VOCATIONS: Lucy is curious about what psychiatrists like her father really do when they are with their patients. Find out more about what one or both of your parents do for a living. You might prepare questions in advance and do a formal interview on tape, or just have an informal conversation. Try to learn something you didn't know before about your parents' vocations. Think about whether you would like to have a similar or different kind of job when you grow up.

FAMILIES: Talk to schoolmates who are the only children in their families and to some who come from families of four children or more. Ask them to discuss the pros and cons of their family size. Assemble the information you collect on a pro/con chart with two headings: Only Child and Large Family. This may help you think about the kind of family you might like to have someday.

PSYCHOLOGY: Do a little research into what psychology is really about. Work as a team and come up with problems for each other to solve.

REFRESHMENTS OR FOOD MENTIONED IN THE BOOK: Make a strawberry pie.

IF YOU LIKED THIS BOOK, TRY...

Lucy Forever, Miss Rosetree, and the Stolen Baby, by Susan Shreve—In this sequel, Lucy's parents adopt a baby, who is kidnapped the day after she arrives at their house. It's Lucy and Rosie to the rescue!

From the Mixed-Up Files of Mrs. Basil E. Frankweiler, by E. L. Konigsburg—Claudia and Jamie run away from home to teach their parents a lesson in appreciation. The brother and sister wind up living in New York's Metropolitan Museum of Art, where they encounter and solve an intriguing mystery (see p. 86).

Some Other Books by Susan Shreve:

The Flunking of Joshua T. Bates

The Gift of the Girl Who Couldn't Hear

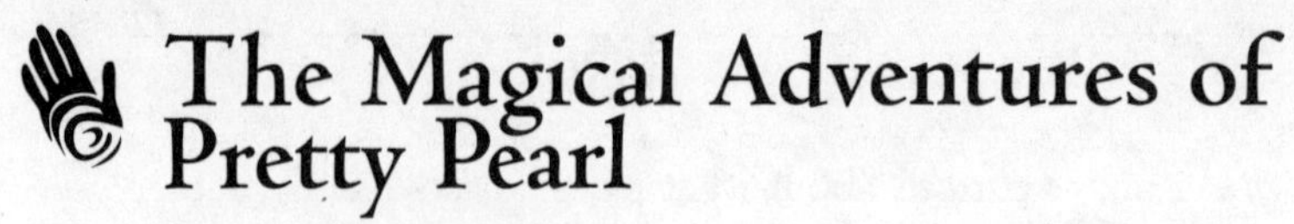

The Magical Adventures of Pretty Pearl

by Virginia Hamilton

I was thrilled to find a fantasy book with an African-American theme.

Pretty Pearl, a god-child high on a mountaintop in Africa, comes down with her Brother, "know-all best god John de Conquer," to sail on a slave ship bound for Savannah, GA. Disguised as a human, Pretty Pearl lives with a band of free blacks who have created their own separate world deep inside a secret forest. Steeped in black history and legend, this powerful fantasy-legend-novel follows Pretty Pearl through centuries of black history, and her transition from god-child to human.

READING TIME: 4–5 hours, about 309 pages
THEMES: slavery, freedom, human behavior, black history

Discussion Questions

- ✦ After giving Pretty Pearl the magic necklace, John de Conquer gives her four rules she must obey: Never promise the necklace to anyone, never take it off for anyone, never use its power to scare anyone, and never use its power to hurt anyone out of spite or anger. In exchange for following these rules, what special powers does the necklace give her? What are your special powers as a human being? Do John de Conquer's rules apply to you? How?

- ✦ Before he and Pretty Pearl board the slave ship, John de Conquer tells Pearl, "Best not interfere or you be come actin' human, too. Be come either a slaver or de enslaved." What is he referring to when he says "actin' human"? Later, he says, "...they never content. They always lookin' to find a way to be above some other ones of them." Do you agree that human beings have a natural urge to possess power over others? Support your opinion with examples from history or your own experience.

- ✦ John de Conquer tells Pearl that humans will always "...twist and turn until they find a worse way to hurt one another." Can you think of examples from history that would lead John de Conquer to say this? Do you believe that human beings might some day improve?
- ✦ Dwahro's "deepest, most solemn wish [is to]...be a man. To be free." Why do you think Dwahro needs to become human in order to be free? For what other reasons might Dwahro want to be a man? When Dwahro finally does become human, what does he have that he didn't have as a spirit?
- ✦ John Henry Roustabout also chooses being human over being a god, even though he will pay for his choice with his life. He tells Pearl, "To be human is about worth de whole world. ..." Review the folk tale and song about John Henry. Why does he choose to be human?
- ✦ When Dwahro reminds Pearl that some white people helped runaway slaves, Pearl says, "It so hard to remember how these humans be." Why do humans confuse Pearl? How are they different from gods?
- ✦ When John de Conquer returns and reminds Pearl that she has broken his rules, she says, "I was just...livin' like a free chile, an' I forgets all about de god chile." How did Pearl behave like a human child rather than a god-child?

ABOUT THE AUTHOR: (see p. 52)

Beyond the Book...

MYTHS AND LEGENDS: The Magical Adventures of Pretty Pearl includes characters and events that make it as much a myth or legend as it is a novel. Read myths from other cultures: Greek, Roman, Norse, Asian, or Native American. Make a compare-and-contrast chart showing how this story is similar to or different from other myths and legends. For example, the Greek myth about Prometheus is also about a character who comes down from a mountain to help human beings, but Prometheus is not a god.

FOLK SONGS: Virginia Hamilton quotes words from several songs and chants that were sung by slaves in the South. Find a book of African-American folk songs that includes both words and music. Learn some of the songs and sing them together. Talk about the meaning the words held for the slaves who first sang these songs.

ART: Think about the picture Dwahro painted on Mother Pearl's apron, and what was so special about it. If you like to draw or paint, try making a picture that, like Dwahro's, goes on forever. For example, you could draw a picture of yourself standing in a room, with a picture on the wall of yourself standing in the same room, and so on.

IF YOU LIKED THIS BOOK, TRY…

Anthony Burns: The Defeat and Triumph of a Fugitive Slave, by Virginia Hamilton—A biography of a slave who escaped to Boston in 1854, was arrested, and whose trial caused a furor between abolitionists and those determined to enforce the Fugitive Slave Acts.

The People Could Fly, by Virginia Hamilton—A treasury of African-American folk tales, stunningly illustrated by Dianne and Leo Dillon.

Second Daughter: The Story of a Slave Girl, by Mildred Pitts Walter—The true story of Elizabeth Freeman, a slave who sued her owner for her freedom in 1781, as told by her fictional sister, Aissa.

Some Other Books by Virginia Hamilton:

The House of Dies Drear (see p. 136)

Cousins (see p. 51)

Zeely

Plain City

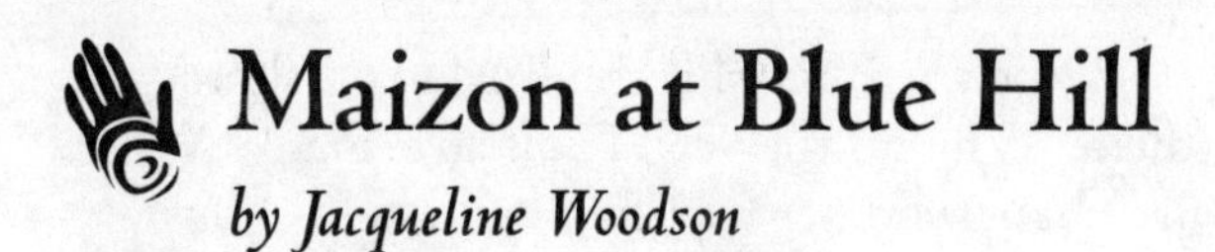

Maizon at Blue Hill

by Jacqueline Woodson

Maizon Singh, a gifted student from Brooklyn, NY, has accepted a scholarship to Blue Hill, a boarding school in Connecticut, where she is one of only five black students. At Blue Hill, Maizon can't figure out where she fits in. She doesn't want to be an "oreo," but sticking with only the other black girls seems wrong, too. At Thanksgiving, Maizon comes home to stay until it's time for college, when she'll be more ready to cope with being on her own.

All three of my children attend schools where they are in the minority, so Maizon's story is a very familiar one. Jacqueline Woodson's work shines a light on cultural identity, acceptance, racism, and elitism. All girls, whether they are in the minority or the majority, can benefit from reading and discussing this book.

READING TIME: 2–3 hours, about 131 pages
THEMES: racism, friendship, identity, decision-making

Discussion Questions

- What is Maizon's main concern about going to Blue Hill? Do you agree that it is important to live in a place where you will find friends of your own race? Why do you think this is, or isn't, important? Have you ever been the minority in a situation with your peers? How did that make you feel?
- What assumptions do Susan, the tour guide, and Sybil, the girl on the debating team, make about Maizon because she is black? What incorrect assumptions does Maizon make about her roommate Sandy because she is white, or about the white girls in her English class when they discuss the book *The Bluest Eye?* Why do you think people *stereotype* others on the basis of race without knowing them as individuals? Have you ever been the object of such stereotyping?

- ✦ Why do the other black girls at Blue Hill think Maizon should be friendly only with them? Why do they call Pauli an "oreo"? What advice would you give about who her friends should be?
- ✦ Charli thinks that black students should go to black colleges so they can get their "education and culture...under one roof." Would you agree if you were one of the black girls at Blue Hill? Whether you would or not, how would you argue your point?
- ✦ Maizon worries about being hurt by prejudice. She wants to be in a place where it is not an issue—where she would not have to think about it. Can you describe a school where prejudice would not be an issue?
- ✦ Marie thinks speaking "correctly" is important, because it "says a lot about who we are." Do you agree with Marie, or do you agree with Maizon that it doesn't matter "as long as you get your point across"? Do you and your friends judge people by the way they speak? Do you think you are judged that way? Do you think it's a good idea to speak as "correctly" as possible in order to be judged more favorably by others?
- ✦ What are Maizon's reasons for not staying at Blue Hill? Do you agree with them? What would you advise her to do, and which arguments would you use to convince her that you were right?

ABOUT THE AUTHOR: Although Jacqueline Woodson now writes full time, she was once a drama therapist for runaways in New York City. Growing up, Ms. Woodson could not identify with the characters she found in "mainstream" literature. When she discovered the books of Toni Morrison, Rosa Guy, and Louise Meriwether, she finally found characters that she could relate to. She says that these authors helped her realize that books could be about anything, and she decided that she would create books that would include characters from a variety of races, ethnic groups, and social classes. Ms. Woodson was born in Ohio, and also spent some time in South Carolina and Brooklyn, where she currently lives. Ms. Woodson enjoys reading, encouraging others to write, sewing, and political debate.

Beyond the Book...

ROLE-PLAYING: Take the parts of Maizon and Sybil, or Maizon and Susan, and role-play a real showdown about racism. Go through the book and make a list of each of their viewpoints to base your dialogue on. After you've gone through the scene once, switch roles. Then talk about your feelings as you played Maizon and Sybil or Susan. What, if any, understanding do you think you gained about race and racism from this activity?

WRITING LETTERS: Maizon has a pile of unmailed letters written to Margaret. She doesn't mail them because she doesn't want Margaret to think she is a failure at Blue Hill. Write one or more of the letters Maizon might have written to Margaret. Then write Margaret's answer(s).

COSTUME: Put together an outfit (including accessories, hair, makeup, whatever you feel is appropriate) that is designed to communicate each of these messages: "I want to fit in," "I want to stand out," "I want to have fun," and "I want to be left alone." Talk about the different ways people present themselves to the world.

GRAPH MAKING: Become more aware of the racial makeup of your own school. While in class, the cafeteria, or the auditorium, make an approximate count of the numbers of students from different races. Show your results on a graph. Then rate the seriousness of racism in your school on a scale of one to ten, with one as the lowest and ten the highest. Write a short paragraph analyzing your results. How do you connect your rating with the figures represented by your graph?

DICTIONARY: Look up the words "prejudice" or "stereotype" in a dictionary. Does each word mean what you thought it did? Talk about situations, besides racial ones, where people have stereotypes or prejudices.

IF YOU LIKED THIS BOOK, TRY...

The Integration of Mary-Larken Thornhill, by Ann Waldron—Because her parents insist, a white northern girl must attend a black junior high school.

The Friends, by Rosa Guy (see p. 82).

Some Other Books by Jacqueline Woodson:

Last Summer with Maizon

Between Madison and Palmetto

Make Lemonade

by Virginia Euwer Wolff

Told in blank verse, this powerful, realistic story depicts the intertwining lives of two poor teenage girls. Determined to save money for college someday, 14-year-old LaVaughn reluctantly takes on a job baby-sitting for the two babies of 17-year-old Jolly, an unwed mother. Though younger by three years, LaVaughn offers patience, understanding, and common sense to help Jolly stay afloat and discover the self-esteem she needs to lift her small family from poverty and a dead-end future.

This book illustrates an interesting twist on what can happen when a baby-sitting job, which most girls look to only as a source of extra money, turns into something more. Sometimes there's a limit to how much "good" one can do.

READING TIME: 2–3 hours, about 200 pages
THEMES: teen pregnancy, self-esteem, goal setting, poverty, education, coming of age

Discussion Questions

- The title of this book comes from the saying: "When life hands you lemons, make lemonade." What do you think this expression means? Discuss examples of ways both LaVaughn and Jolly make lemonade from the many lemons in their lives. What qualities do you think people must have in order to transform their misfortunes into positive action?
- Both LaVaughn and Jolly are teenagers born into very poor circumstances, yet they are very different. Compare the two girls.
- Discuss why you think LaVaughn has such strong self-esteem and Jolly's self-esteem is so low. What kinds of experiences build up self-esteem? What tears it down?

- LaVaughn says: "Homework is a completely required thing like a vaccination." What do you think LaVaughn means by this? If homework is like a vaccination, what disease does it prevent?

- What makes LaVaughn decide to take the baby-sitting job? Discuss her baby-sitting experiences and how she handles them. If you sometimes baby-sit, talk about some of the experiences you've had that are similar to those in this book. Did you learn anything about child care from this book?

- Describe LaVaughn's mother. Describe the mother-daughter relationship they have. Discuss some ways your own mother-daughter relationship is similar to or different from the one depicted in this book.

- What kind of a parent do you think Jolly is? Do you think a 17-year-old can be a responsible parent?

- At the beginning of chapter 6, LaVaughn and her mother have a conversation about a loaded subject—whether LaVaughn can have a job and keep up her grades. Reread the first page of chapter 6 aloud. In what ways is that conversation similar to those you've had with your own mother or daughter? Daughters: When you want to ask your parents for something you know they're not going to like, how do you plan your request? Mothers: When you know your daughter is going to bring up something unpleasant, how do you plan to voice your concerns?

- Do you think LaVaughn is too involved with Jolly's life? What would you have done in her situation?

- Why does LaVaughn's baby-sitting job become so important to her even after it starts to interfere with school?

- Describe LaVaughn's vision of her future. Do you think she has the determination to reach her goals?

- LaVaughn's mother voices a belief many people have: "Some people make a bad bed. They just have to lie in it." What do you think of this opinion? LaVaughn does something wonderful shortly after hearing her mother say this: she teaches Jeremy how to make a neat bed. What does this tell you about LaVaughn?

- The book is open-ended—not every problem LaVaughn and Jolly have is solved. What do you think the future holds for these girls?

ABOUT THE AUTHOR: Virginia Euwer Wolff (often known as Jinny) always wanted to be a writer, but did not write her first novel until she was almost 40. Since then she has written two other books. Before beginning to write, Ms. Wolff taught in New York and Philadelphia. A Portland, OR, native, she eventually returned to Oregon to live in Oregon City.

Beyond the Book...

SELF-ESTEEM: In LaVaughn's self-esteem class, the "Steam Class," the students boost their self-confidence by reciting challenges they are capable of overcoming. Before your book discussion, prepare a list of completed sentences that begin with the phrase: "I am capable of. ..." Then recite your "capables" to each other.

TALKING TO MOMS: A couple of times during the book, LaVaughn has to broach issues with her mom that she knows her mom isn't going to want to hear. I think all girls have a way of trying to smooth the waters when they want something from their moms or need to tell them something unpleasant. My mom could always tell when I wanted something because she would come home and *all* my chores would be done and the house would be spotless. I remember being so frustrated when the first words out of her mouth were not, "How nice," but, "What do you want?" Talk about how you approach—or approached—your moms in situations like this. Is this kind of technique usually effective? Why or why not?

REFRESHMENTS OR FOOD MENTIONED IN THE BOOK: Make lemonade! Prepare and serve real lemonade made from lemon and ice and sugar dissolved in water. While enjoying the lemonade talk about how you have transformed difficult life experiences into positive ones.

IF YOU LIKED THIS BOOK, TRY...

The Mozart Season, by Virginia Euwer Wolff—This 1991 novel by the same author, and an ALA Notable book, portrays a teenage girl with a strong vision of her future as a musician.

Detour for Emmy, by Marilyn Reynolds—This novel is also about a teenage girl dealing with unwed motherhood and a dysfunctional family.

Some Other Books by Virginia Euwer Wolff:

Probably Still Nick Swanson

Bat 6

Maniac Magee

Jerry Spinelli

This fast-paced, almost-tall-tale story tells of a 12-year-old orphan, Jeffrey Lionel Magee, nicknamed Maniac Magee for his whirlwind energy. Maniac becomes a legend when he blows into a small Pennsylvania town looking for a place to stay. No one has ever seen someone with Maniac's unique combination of athletic prowess, fearlessness, street smarts, and open acceptance of everyone he meets—old, young, black, white, rich, or poor. Though humorously told, readers can expect some heartbreaking moments in the story as well.

Many years ago, I saw a play written and performed by a group of homeless people. The short monologues, mostly about their experiences, left no dry eyes—and surely not one audience member left there unmoved. I think this book has a similar effect on its readers. Its portrayal of a homeless young boy is touching and consciousness-raising for us all.

READING TIME: 2–3 hours, about 184 pages
THEMES: homelessness, attachment, acceptance, literacy, race, friendship

Discussion Questions

- ✦ What is a tall tale? Why does this book fit that description?
- ✦ Maniac has had a painful, loveless childhood. Where do you think his strength, spirit, and ability to love come from?
- ✦ Can you think of difficulties in your own life that wound up making you a stronger person?
- ✦ Why is Maniac so appealing to so many people? Do you find him appealing? Why or why not?

✦ Maniac infuriates some of the bullies in the town of Two Mills. Why? Have you ever been bullied or been a bully yourself?

✦ When Maniac finally gets a home with the Beales family, he chooses to sleep on the floor. "Maniac just couldn't stand being too comfortable." What does it mean to be "too comfortable"? Why can't Maniac stand it?

✦ Why do address numbers on the fronts of houses mean so much to Maniac?

✦ Certain sounds of the Beale household and lifestyle give Maniac a sense of home and belonging: pancake batter hissing on the grill, gospel music, the noisy Fourth-of-July block party. What sounds give you a feeling of belonging to your family and community?

✦ "Maniac loved almost everything about his new life. But everything did not love him back." What are the things Maniac loves about his new life in Two Mills, PA? What are the things that don't love him back?

✦ Maniac has a kind of innocence and goodness that help him recognize and accept people without judging them by status or race. How does that help or hurt him? Why do you think his blindness to social or racial labels sometimes gets him in trouble?

✦ After Maniac meets Grayson, the old man who is down on his luck, why won't Maniac go to school? Why does Grayson want him to go to school?

✦ The author maintains a humorous style throughout the book, even when dealing with the painful subjects of homelessness, bullying, racism, and death. How did Spinelli's humorous style affect you during the scenes when Maniac is bullied, when he has to leave the Beale family, when Grayson dies?

ABOUT THE AUTHOR: Jerry Spinelli has been a writer all of his life. His childhood in Norristown, PA, was the inspiration for his book *Maniac Magee,* which won the Newbery Medal in 1991. His books are known for their honesty and comedy. Many of the characters and situations in his books are drawn from people in his own life. He now lives in Phoenixville, PA, with his wife, Eileen, who is also an author, and two of their six children.

Beyond the Book...

READ ALOUD: As part of the book discussion of *Maniac Magee,* read aloud *The Little Engine That Could* and *Mike Mulligan and the Steam Shovel.* After the reading, discuss why Spinelli chose these two particular books for Maniac to use for teaching Grayson to read. What is the message of these books? How does that message apply to Maniac and to Grayson?

LITERACY: The subject of illiteracy comes up when Maniac discovers that Grayson can't read and decides to teach him. Since you obviously love reading enough to spend free time reading and discussing books, share with each other why you care about reading so much. Imagine what daily life must be like for adults who can't read. Is it just books illiterate adults are deprived of, or something more?

VOLUNTEER: Brainstorm ways to combat illiteracy, as a group or individually. Perhaps you can volunteer time to help teach children reading in the schools or in after-school programs. Train as literacy volunteers at the library, volunteer time to work at a library, or donate good books to schools, day care centers, libraries, or homeless shelters.

REFRESHMENTS OR FOOD MENTIONED IN THE BOOK: Maniac loves Mrs. Beale's home cooking—her pancakes, meatloaf, and coconut pie are some examples. For an Eagle Scout project my son, Leroy, cooked and served dinner at a homeless shelter. He even arranged for some of the scouts to play music during the meal. As a group or individually, you might consider organizing, preparing, and delivering a home-cooked meal—potluck style—to a local charity.

IF YOU LIKED THIS BOOK, TRY...

The Great Gilly Hopkins, by Katherine Paterson—Gilly Hopkins is another lively white orphan who finds her true home with a memorable black matriarch (see p. 113).

Foster Mary, by Celia Strang—Four unrelated foster children learn to adjust to the routines of family life.

Some Other Books by Jerry Spinelli:

Crash

Jason and Marceline

The Moon Over Crete

by Jyotsna Sreenivasan

For Lily, a suburban, 11-year-old girl, the trials and tribulations of growing up are often overwhelming, despite sympathetic and enlightened parents. Her time-travel adventures to ancient Crete (where men and women enjoyed true equality) under the guidance of her music teacher, offer Lily new possibilities and hope for her future.

I work hard to pass on my belief that girls can do anything to my daughters, though I never want them to have to be "super women." The key for our girls is balance. Lily's journey offers a wonderful opportunity to discuss gender roles and finding one's place.

READING TIME: 1–2 hours, about 124 pages
THEMES: sexism, race, power, family, relationships, friendship

Book Discussion Questions

✦ Lily was very upset by her encounter with Chuck at school. Why was that episode so upsetting to her? How would you feel if that happened to you? Have you ever been in a situation that made you uncomfortable in that way? How did you handle it?

✦ How does Lily feel about her changing relationship with her best friend, Lauren? Have you had friendships that changed over time? How did you feel?

✦ When Geeta is concerned that Chuck is teasing Lily because she's Indian-American, Lily says, "Who cares about race anyway?" Why would Geeta think Lily was being teased because of her race? Do you agree that no one cares about race? Have you ever been teased because of your race?

✦ Lily's flute teacher, Mrs. Zinn, observes that "It's hard nowadays being a girl." What does she mean? Do you agree? Discuss whether you think it was easier being a girl in other eras, and if so, give your reasons.

- ✦ When Lily first arrives in Crete, she can't tell whether the children she sees playing are girls or boys. How does she react? How would you respond?
- ✦ How is Mashi's relationship with her father different from Lily's with her father? What kind of relationship do you have with your father? Did reading this book make you want to take another look at that relationship?
- ✦ How is the Cretans' sense—and practice of—equality different from Lily's community at home? How does that affect Lily, and change her sense of the way the world works?
- ✦ When Lily discovers that physical appearances don't matter in Crete like they do at home, how does she respond? How would you react?
- ✦ How does Lily feel when she finds out that she can talk back to her dancing teacher in Crete? How is it different from her experience at home?
- ✦ Lily feels important in Crete. Why? Is there something you do, or an experience you've had, that makes you feel valued in that way?
- ✦ Why does Lily want to remain in Crete? What would you have wanted to do?
- ✦ In what ways does her time in Crete change Lily? What kind of adult do you think Lily will become as a result of her experiences there?

ABOUT THE AUTHOR: Jyotsna Sreenivasan was born in Ohio, moved to India, and returned to the United States all before she turned seven. She is an Indian-American whose books are often composites of her childhood memories and experiences. She feels that her ethnic background plays a large role in her life, and therefore her books are often about characters who are also influenced by their backgrounds and ethnicity. She began writing almost as soon as she learned to read and eventually published several books.

Beyond the Book...

GREEK MYTHOLOGY: Research ancient Greece, especially Crete, and Greek mythology, with a particular focus on rituals pertaining to the celebration of women and goddesses (Aphrodite, Athena, Helena, and so on). Discuss what it would have been like to live in a culture and society where goddess-worship existed and women were valued to that extent. What do you think the benefits would have been? Can you think of any disadvantages?

CELEBRATION: Prepare your own cultural celebration, complete with dancing and displaying crafts you are especially proud of. Don't feel limited by the book's description of what happened in Crete—open the possibilities to milestones and rituals that you find meaningful.

UTOPIA: Imagine yourself transported to a gender-equal utopia, like the Cretan society described here. What would be important elements of that society? How would you envision and create a perfect world?

REFRESHMENTS OR FOOD MENTIONED IN THE BOOK: Provide a Cretan feast, with olives, feta cheese, and fresh pears. Talk about the kinds of food that would have been eaten in daily life in the ancient world—like plain yogurt, honeycombs, or simple bread—and what made festival foods so important.

IF YOU LIKED THIS BOOK, TRY...

A Wrinkle In Time, by Madeleine L'Engle—A spirited heroine has to travel through another dimension to accomplish an important mission (see p. 314).

Some Other Books by Jyotsna Sreenivasan:

Aruna's Journey

The Moorchild

by Eloise McGraw

In this haunting and evocative fantasy, a young girl who is half human, half fairy struggles to find a place for herself in an unwelcoming world. Her ultimate acknowledgment of her unique identity helps her accept her destiny and attain a surer sense of who she is.

I think we all have times in our lives when we feel out of place. When I was young I didn't like going to parties because I couldn't dance. The other kids teased me. It was expected that all African-American kids dance, and I felt like I was the only one who couldn't. My mom kept sending me to parties and I never learned. But I did learn not to care what other people thought. Now I still don't know how to dance, but I sure love dancin' the night away.

READING TIME: 2–3 hours, about 241 pages
THEMES: prejudice, responsibility, trust, family, loneliness, friendship, self-discovery, individuality, redemption, loss, memory

Discussion Questions

- ✦ Why doesn't Moql (known in the human world as Saaski) have a relationship with her mother, who is a full-fledged member of the Folk? Imagine how it would feel to be raised the way Folk children are reared. Would you like it or not ? Explain your reasons.
- ✦ In the Folk world, "it was a life without yesterdays or tomorrows." What would be good about living like that? What would be bad?
- ✦ Moql, as a half-human, half-Folk child, experiences emotions and feelings that the other Folk don't. How does that make her feel? In what ways is she like a human? In what ways is she more like a Folk? How do these conflicts in her personality affect Moql/Saaski? What does it mean for her to be neither fully one nor the other?

- Being different, yet having to conform to community expectations, makes Saaski feel trapped. What expectations make her feel that way? Have you ever been in a situation that made you feel that way?
- Why does Saaski always feel lonely? What changes in her life help her feel less lonely?
- How is Saaski's friendship with Tam important?
- How are Old Bess and Saaski alike? How are they different? Why is their relationship important to each of them?
- The other villagers use Saaski as a scapegoat when their children become ill. What does it mean to be singled out like that? How does Saaski respond?
- Her parents and the villagers see the moor one way. Saaski sees it quite differently. How do these dissimilar perceptions of the moor help Saaski uncover her true identity?
- Why is finding the set of bagpipes a turning point in Saaski's life?
- What's significant about the gift Saaski ultimately leaves with her foster human parents? How does it reflect Saaski's character?

ABOUT THE AUTHOR: Newbery Honor winner Eloise Jarvis McGraw was born on December 9, 1915, in Houston, TX, and began writing when she was eight years old. She is also an artist, dancer, actress, and director, and enjoys horseback riding and studying ancient Egyptian history. She is married to the writer William Corbin McGraw.

Beyond the Book...

MYTHOLOGY: Research the myths and legends that surround fairies, gnomes, elves, and other Folk-like figures. Focus particularly on stories about changelings, and customs and practices of these otherworldly creatures.

HERBAL REMEDIES: Herbal treatments and knowledge of the kind that Old Bess possesses and Saaski learns are coming back into popularity. Have the mother-daughter pairs pick one or two herbal remedies to present to the group, and have each pair bring in the appropriate herbs to show to the group. If members of the group remember any nonmedical grandmother remedies (including, but not limited to, chicken soup and broth for colds, or compresses for congestion), have them share these stories with the others.

REFRESHMENTS OR FOOD MENTIONED IN THE BOOK: During your discussion, play traditional Celtic music, including flutes and bagpipes, to suggest the atmosphere of the book. Serve herbal teas, simple brown bread, and vegetable soup as refreshment, in keeping with the rural lifestyle described here.

IF YOU LIKED THIS BOOK, TRY...

The Princess and the Goblin; At the Back of the North Wind, by George MacDonald—These stories deal with encounters between humans and nonhumans.

A Midsummer Night's Dream, by William Shakespeare—This play is best for slightly older readers.

Some Other Books by Eloise McGraw:

The Golden Goblet

Mara, Daughter of the Nile

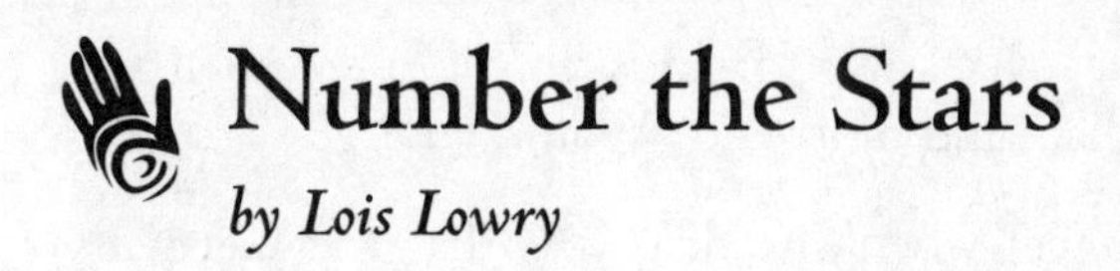

Number the Stars

by Lois Lowry

Annemarie Johansen, a young Danish girl growing up during the Nazi occupation, is forced to summon up all her courage to help her best friend, a young Jewish girl. In the process, she discovers strengths that she didn't know she possessed which forever alter her self-image. Even younger readers will enjoy this accessible novel about World War II and the Danish Resistance's efforts to smuggle the Jewish population to safety.

Lois Lowry came to Morgan's school this year, after her class had read *The Giver,* one of the author's other books. During her visit, Ms. Lowry read from several of her books and told stories about each one. Morgan particularly likes *Number the Stars* because, she says, "It is interesting to read about how other girls lived, what they went through, and how hard it was for them."

READING TIME: 1–2 hours, about 137 pages
THEMES: courage, individual versus community values, friendship, secrecy, integrity

Book Discussion Questions

- What kind of friendship do Annemarie and Ellen have? How can you tell? Why are they such close friends? Do you have a friend you are that close to? What's important about that relationship for you?
- How does the encounter with the Nazi soldiers on the way home from school affect Annemarie? How would it make you feel to have foreign soldiers watching your behavior?
- What does Mrs. Rosen mean when she tells the girls, "It is important to be one of the crowd, always"?
- When Annemarie realizes that Ellen's family is Jewish and is threatened by the Nazi presence, her mother tells her that they will pro-

tect the Rosens because "that's what friends do." Does Annemarie feel capable of being that kind of friend? Would you?

- ✦ How does Annemarie feel when she discovers that the Germans have closed Mrs. Hirsch's button shop, simply because the new laws forbid Jews from running businesses? What does she mean when she says, "Now I think all of Denmark must be bodyguard for the Jews?" How do you think you would have responded in that situation? Why?
- ✦ How does her parents' disappearance affect Ellen? How must it feel to have to pretend to be someone you're not?
- ✦ Playing pretend is a normal part of growing up. Annemarie and Ellen have to pretend that they are really sisters to fool the Germans. How is this different than games of make-believe?
- ✦ Right before the soldiers come in to her bedroom, Annemarie grabs the necklace with the Jewish star off Ellen's neck and holds it so tightly that it leaves an imprint on her hand. What is significant about this gesture? How is that imprint representative of how Annemarie is affected by what's going on?
- ✦ Why are the seashells Peter gives Annemarie so important to her? Have you ever gotten a present that was as important to you? What's so significant about it?
- ✦ Sometimes the adults around Annemarie and Ellen don't tell the truth. Why is it all right for them to lie? What's dangerous about the truth in their situation? How does the war change people's ideas of what's right and wrong?
- ✦ Why does Annemarie's family let her believe that she has a Great-Aunt Birte? How does Annemarie feel about the secret that her family has kept from her? How does that affect her? How would you have felt? Why does Annemarie have to lie to Ellen? Can you foresee any circumstances or situations where you might have to, or want to, do the same thing?
- ✦ What gives Annemarie the courage to deliver the lunch basket to her Uncle Henricks? What do you think you would have done had you been in her place?
- ✦ Annemarie recognizes that even without their books, furniture, and home, Ellen and her family still have their pride. What does it

mean to have pride? Where does Annemarie think that pride comes from? What do you think?

- ✦ Were you surprised to discover that Annemarie's sister had been a part of the Danish Resistance movement? Peter?
- ✦ What does "Number the Stars" mean?
- ✦ Usually people think of bravery and courage as traits possessed by heroes, warriors, or exceptional individuals. Annemarie, her family, and Peter are brave, each in their own way. How do Annemarie and her family show it? How does Peter?
- ✦ Not only do Annemarie, her family, and Peter take courageous chances for close friends, but they risk their lives for strangers. What does courage mean to you? What kinds of risks would you be willing to take for your friends? For strangers?

ABOUT THE AUTHOR: (see p. 98)

Beyond the Book...

QUILTING: In the book, Annemarie's grandmother has a quilt she stitched. Find some fabric you like, cut it into squares, and sew the pieces together as a starter for your own patchwork quilt. Talk about how handmade things like this (or the handkerchief Annemarie delivers to Uncle Henriks) can come to be so valuable within a family. Find some things that are important in your family and talk about why they mean so much.

DENMARK: Go to the library and look for other books about the Danish resistance to the Nazis. Find facts like how many Jews there were in Denmark and how that population was affected by the Holocaust. Look on a map to locate where in Denmark Annemarie and Ellen lived.

FLASHLIGHT READING: Find an area in your home where you can create a hidden living space. Go into the space and spend some time there—with a flashlight and the book—and read over the passages where Annemarie and Ellen read in bed that way.

MOVIE: Rent *Anne Frank: Diary of a Young Girl* to get a sense of what it was like to live in hiding under Nazi occupation.

REFRESHMENTS OR FOOD MENTIONED IN THE BOOK: Make homemade applesauce and serve it with roast chicken and potatoes like Annemarie's family had in the book. Or, for a snack, make oatmeal and herbal tea.

IF YOU LIKED THIS BOOK, TRY...

Behind the Bedroom Wall, by Laura E. Williams—This book deals with similar issues and themes about hiding a Jewish family.

Anna Is Still Here, by Ida Vos—This story explores what happens after a Jewish girl is hidden during the war and survives to be reunited with her family.

Anne Frank: Diary of a Young Girl, by Anne Frank—The journal entries of a young girl describe a life in hiding during the war.

Some Other Books by Lois Lowry:

All About Sam

Anastasia Krupnik

Find a Stranger, Say Good-bye

The Giver

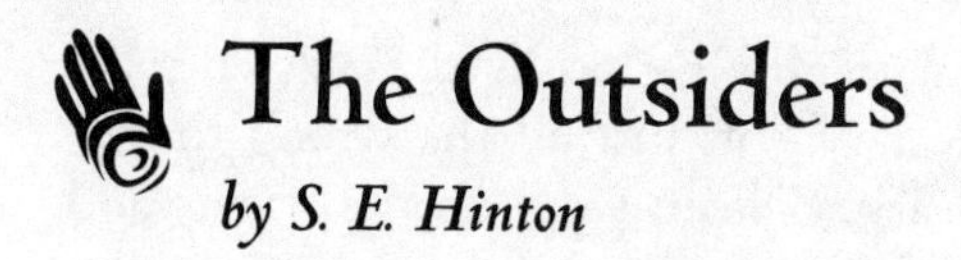

The Outsiders

by S. E. Hinton

In this enduring novel, written by S. E. Hinton when she was 16, two rival high school gangs fight for power. Bored and aimless, the wealthy and advantaged Socs constantly provoke the local "greaser" kids, one of whom, Ponyboy, narrates the story. The greasers have little left to lose but their last scraps of pride. This action-packed novel portrays young lives circumscribed by poverty, death, abuse, and alcoholism.

READING TIME: 2 hours, about 180 pages
THEMES: acceptance and rejection, violence, siblings, loyalty, friendship, death, coming of age

Discussion Questions

- The author, S. E. Hinton, was 16 in 1967, when she wrote *The Outsiders.* Can you tell that a teenager wrote this book? How? Do you think a talented 16-year-old could write a book with similar themes and characters today?
- Describe what various groups have in common in your current school or the schools you once attended. Are there or were there conflicts among the groups similar to those in *The Outsiders*? How do you—or did you—fit into these groups?
- Ponyboy resents that his older brother, Darry, rides him hard about school, friends, and responsibilities. Why is Darry much harder on Pony than on Sodapop? Do you think that Darry's high expectations for Pony are an expression of love or power?
- Cherry, one of the girls in the Socs, tells Ponyboy: "We're sophisticated—cool to the point of not feeling anything." What do you think of this definition of being "cool"? What is your own definition of "cool"?
- Why does the Robert Frost line, "Nothing gold can stay" mean so much to Ponyboy? What do you think this line means?

- ✦ Ponyboy has this insight about the life he has lived: "Sixteen years on the streets and you can learn a lot. Just all the wrong things, not the things you want to learn. Sixteen years on the streets and you see a lot. But all the wrong sights, not the sights you want to see." What are the things Pony wants to learn and see besides what he has experienced? In your own life, what do you hope to learn and see that you haven't experienced yet?
- ✦ There are no adult main characters in *The Outsiders.* The teen characters experience passion, love, hate, and violence amongst themselves and completely apart from the world of adults. Mothers: Do you remember undergoing major experiences and feelings apart from your parents when you were a teenager? Share your experiences with your daughters.
- ✦ Loyalty and friendship are the glue that binds the characters in *The Outsiders* together. How do the characters show loyalty to their friends? How do your friends show loyalty to each other?
- ✦ Ponyboy's intelligence and sensitivity shine through all his experiences. Speculate on what kind of future you imagine he might have as an adult.

ABOUT THE AUTHOR: Susan Eloise (S. E.) Hinton was a teenager when her first book, *The Outsiders,* was published. Her books are extremely realistic and offer a view of young-adult life that does not sugarcoat the struggles that an adolescent faces every day. Since the main character of her first book was a young man, Hinton's publishers advised her to publish under the name S. E. Hinton. They wanted to ensure that male readers did not disregard the book because of its female author. Since then, she has written a number of other novels; many have been turned into movies. Hinton lives with her husband and son in Brooklyn, NY, where she enjoys taking classes at the local university, being involved in her son's school, horseback riding, and cooking.

Beyond the Book...

DIALOGUE: Although the novel is told in Ponyboy's voice, other characters come alive in the many dialogue-filled scenes that convey the pain, feelings, and thoughts of troubled teenagers. Choose a scene to read aloud to each other. Write your own dialogue about a gathering of your friends and compare it to those in the book.

MOVIE: View the two movies featuring Hintons' characters: the 1983 Francis Ford Coppola movie, *The Outsiders,* and the 1985 film *That Was Then, This is Now,* based on Hinton's sequel to *The Outsiders.*

REFRESHMENTS OR FOOD MENTIONED IN THE BOOK: The boys in this book eat a lot of junk food. Go to a fast-food restaurant and take out lunch, or serve chocolate cake for breakfast. For a healthier meal, make egg, bacon, and tomato sandwiches—with some chocolate milk on the side.

IF YOU LIKED THIS BOOK, TRY...

Rumblefish; That Was Then, This is Now, by S. E. Hinton—These two novels are sequels to *The Outsiders.*

Some Other Books by S. E. Hinton:

The Puppy Sisters

Tex

The Poison Place: A Novel

by Mary E. Lyons

Intrigue, deception, and danger mark the existence of Moses Williams, a slave in the household of a Philadelphia gentleman (and amateur scientist) during the Revolutionary War era. In this book, he tells his daughter the compelling story of his survival, his struggle to earn his freedom and maintain his dignity.

READING TIME: 2–3 hours, about 155 pages
THEMES: trust, betrayal, racism, slavery, sibling rivalry, self-esteem

Discussion Questions

✦ Moses Williams tells his daughter, "There's other kinds of bondage. One is long words on fancy paper that says a white person can own a black one. T'other plays peekaboo at the back of the mind, so's a fellow hardly knows it's there." What does he mean? Have you ever experienced the latter kind?

✦ What kind of relationship does Moses have with his master, Mr. Peale? With Mr. Peale's sons? How do those relationships, especially the one with Raphaelle Peale, change during the course of the book?

✦ Peale claimed that his slaves were like "family." What did he mean by that? How did Moses and his family perceive that claim? Did it seem true from your perspective?

Moses' experiences in this book reminded me of the stories my grandfather used to tell me.

✦ How does Moses' family feel about being kept in slavery when other Philadelphia slaves are being set free? How does it affect their feelings for Mr. Peale?

- ✦ Moses' feelings about Rafe, his master's son, are complicated. Have you ever experienced similar ambivalence about a friend? If so, describe the situation and what happened.
- ✦ What kind of relationship does Moses have with his parents? How does slavery influence that?
- ✦ Why do you think Mr. Peale keeps sending Rafe to do the dirty—and dangerous—job of acquiring specimens for his museum and working with arsenic to preserve them? Why is Rafe so willing to obey his father and try to please him, even though he treats Rafe so poorly?
- ✦ Moses observes that Rafe "felt like he couldn't do anything right." Why does he feel that way? How do those feelings influence the choices and decisions he makes? Have you ever felt that way?
- ✦ Moses wasn't given much education because he was a slave, yet his understanding and intuition about the workings of the human heart surpassed that of the Peales'. Discuss important knowledge or lessons you've acquired that had little to do with formal or academic learning, and what was significant about them.
- ✦ One of the things Moses says to his daughter is: "a lie can grow like mold on bread." What does that mean in the context of the book? Have you seen examples of the truth of that saying in your own life? If so, describe what happened.

ABOUT THE AUTHOR: Mary E. Lyons was born and raised in the South, and many of her books concern the South. Lyons moved around a lot in her youth and often sought refuge in books. She comes from a family of avid readers and her father always made sure that she and her siblings visited the local library often. Ms. Lyons eventually became a school librarian and reading teacher. Noticing that her students enjoyed the stories created by African-American and women writers she set out to educate them about these writers. Her first book was *Sorrow's Kitchen: The Life and Folklore of Zora Neale Hurston.* She has since written numerous books that she hopes will carry on the lessons she sought to teach in the classroom. Ms. Lyons and her husband live in Charlottesville, VA, where she plays the banjo and penny whistle for a band called the Chicken Heads.

Beyond the Book...

SLAVERY: Gather some background material on slavery during the colonial era, when it wasn't necessarily limited to Southern plantations. Discuss how slaves contributed to the development of the young United States through their work in business and other enterprises.

SILHOUETTES: Moses' skill at silhouette cutting helped rescue him from the worst work in the museum. Find books about silhouette cutting, its origins and practices, and see if you can try to make silhouettes of one another.

NATURE MUSEUM: Naturalistic dioramas play a crucial role in this book. Arrange to visit a museum exhibit, and discuss the process and effort that went into producing them.

LEGACY BOXES: Moses bequeaths his daughter, Mag, a box that has meaningful objects and documents. Construct a box and decide what to put into it, as a legacy. What would you include? Talk about your choices, and why you selected what you did.

MOVIE: See and discuss the movie *Amistad,* directed by Steven Spielberg. This film deals with a lot of issues of slavery and freedom.

REFRESHMENTS OR FOOD MENTIONED IN THE BOOK: As a snack during your discussion, share Irish soda bread and tea. If there is time, bake the bread together.

IF YOU LIKED THIS BOOK, TRY...

Roll of Thunder, Hear My Cry, by Mildred D. Taylor—A nine-year-old and her family face racial prejudice in the Depression-era South (see p. 212).

Steal Away, by Jennifer Armstrong—A white orphan and a black slave disguise themselves as boys and escape to the North in 1855.

Ramona the Pest

by Beverly Cleary

The growing pains and pleasures of childhood "firsts" are the subject of this warm and funny novel—first day of school, first lost tooth, first pair of new boots, crossing the street alone for the first time. Through it all, Ramona Quimby, the outspoken "baby" of the family, tackles these experiences with a determination that only seems pesty.

READING TIME: 2–3 hours, about 192 pages
THEMES: coming of age, friendship, family

Discussion Questions

- ✦ At times, family members and friends think of Ramona as a pest. Ramona, however, prefers to think of herself as "...a girl who could not wait. Life was so interesting she had to find out what happened next." Do you think being "pesty" is annoying or a sign of enthusiasm for life?
- ✦ Share memories of your own first days of kindergarten. How were your kindergarten experiences similar to or different from Ramona's?
- ✦ Ramona says she can't understand how adults can believe that growing up goes by so fast. Discuss whether or not you think growing up takes place slowly or quickly. How has your opinion about the "speed" of childhood time changed as you've grown older?
- ✦ Ramona looks forward to losing a tooth, riding a bike, and wearing lipstick. What were some of the things you looked forward to when you were in kindergarten? What are some of the milestones you look forward to now?
- ✦ Ramona secretly loves being her mother's "baby," though she doesn't want to be treated like one. What is your position in your own

family—first, middle, youngest? What are the advantages and disadvantages of your place in your family's birth order?

- The adults in Ramona's life—her parents, her teachers, people in the neighborhood—frequently misunderstand what is important to Ramona. Daughters: What do you think adults misunderstand about what it's like to be your age? Mothers: What do children have trouble understanding about what it's like to be a parent?
- Ramona hates wearing hand-me-down boots and cannot wait to get brand-new boots. Describe the feeling of getting a wonderful new pair of shoes.
- Ramona can hardly wait to be a scary Halloween witch. She is a little disappointed when she sees so many other witches running around. Why do you think so many girls want to be witches at Halloween?
- Ramona is both brave and terrified about not being recognized in her witch costume. Share any experiences you've had of being lost in a large crowd at a young age.
- For Ramona, losing her first tooth is a glorious experience. What kind of experience was it for you?
- What makes someone a pest? Do you think Ramona Quimby is a pest or not?

ABOUT THE AUTHOR: (see p. 91)

Beyond the Book. . .

SHOW AND TELL: Ramona loves the Show and Tell event in her kindergarten. Set aside a Show and Tell time during your book discussion. Find a favorite possession to show and talk about with each other. Talk about how and why certain objects become favorites.

PICTURES: Bring nursery school and kindergarten pictures or writings from those grades to share during the book discussion. Photos, artwork, or early writing efforts make wonderful discussion starters. Try to re-create your early kindergarten days—the confusion, setbacks, triumphs, embarrassments. Talk about how these school experiences were similar to Ramona's.

CONTEST: Create a New Sandwich Contest. Bring your entry to the book club meeting. Have a tasting session, with a preselected panel of judges.

REFRESHMENTS OR FOOD MENTIONED IN THE BOOK: Soup and sandwiches are comfort foods in *Ramona the Pest.* Talk about the meaning of the phrase "comfort food," and ask book discussion members to prepare samples of comfort foods from their own families to share with the book group. Invite members to talk about how certain foods became comfort foods in their families.

IF YOU LIKED THIS BOOK, TRY...

Beezus and Ramona, by Beverly Cleary—More of the Quimby family, this time focusing on Ramona's relationship with her sister.

Henry Huggins, by Beverly Cleary—This book is about Ramona's savior and hero, Henry Huggins.

The Not-Just-Anybody Family, by Betsy Byars—This book is warm and funny, in the same spirit of the Ramona books.

Amber Brown Is Not a Crayon, by Paula Danziger—A third-grader loses her best friend and copes in the classroom.

Some Other Books by Beverly Cleary:

Dear Mr. Henshaw
Ellen Tebbits
The Mouse and the Motorcycle
Fifteen

Red Scarf Girl: A Memoir of the Cultural Revolution

by Ji-Li Jiang

Ji-Li Jiang tells of her own growing up in China during the Cultural Revolution. As descendants of a landlord, Ji-Li and her family are under constant threat from the government, and live in fear of arrest. Finally, with the arrest of her father, Ji-Li is faced with making a decision to betray her family or sacrifice her own future in the Communist Party. The book ends with her courageous decision. In an epilogue, the author tells how she and her family left China to make a new life in the United States.

When I met Ji-Li Jiang at a book conference in California, I had not yet read her memoir. If I could thank her now, I would, for sharing her extraordinary story with us in this book.

READING TIME: 3–4 hours, about 285 pages
THEMES: patriotism, family, loyalty, courage, friendship, choice

Discussion Questions

- The "Four Olds"were: old ideas, old culture, old customs, and old habits. Why do you think Mao wanted people to "destroy the Four Olds"? What are some of our "olds" in the United States, or some "olds" in your family? How would life be different without them?
- In the United States and other capitalist nations, some people are rich while others are poor; some live in luxury, while others do not have the basic necessities. Communism tries to solve this problem by elevating the "common people" and lowering the status of the rich and powerful. According to the author of this book, how well did this solution work to make people more equal? Do you think

all people should be equal in this way? Do you have any ideas for narrowing the gap between rich and poor under a democratic system?

- The United States has a constitution that protects the rights of individuals—people's *civil liberties* or *civil rights.* What rights that we take for granted were not protected in China during the Cultural Revolution? One example is freedom of speech, freedom to say what we think without fear of being punished, even if we criticize our own government. Do you think that people should be allowed to speak publicly in support of destructive ideas such as racial or religious prejudice? What could happen if their freedom of speech were restricted? Talk about problems with restricting freedom of the press and the media as well.

- Ji-Li Jiang tells of many terrible experiences she, her family, and friends lived through during the Cultural Revolution. Which of these seem the worst to you? What makes these experiences so terrifying and shocking? What purpose did treating people so horribly serve in Mao's China?

- Although Ji-Li loves her family deeply, she is sometimes tempted to break with them, as when Teacher Zhang suggests that she be a guide at the Class Education Exhibition. She tells herself that she must love Chairman Mao and become an "educable child." At one point, she even yells, "I hate this landlord family," and soon after, considers changing her name. Do you blame Ji-Li for behaving this way? Why, or why not? Would you have blamed her if she had separated herself from her family or betrayed her father? Discuss the reasons for your answer.

- Do you think the equivalent of China's Cultural Revolution could ever happen here? If you do think so, talk about how it could happen; if you don't think so, talk about what in our government or society would prevent a leader like Mao from coming into power in the United States.

ABOUT THE AUTHOR: Ji-Li Jiang's first book, *Red Scarf Girl: A Memoir of the Cultural Revolution,* is the autobiographical account of the author's life in China while Chairman Mao was in power. In this book, Ms. Jiang tried to re-create the emotions she felt as a child and to portray the struggles that she and her family endured. After Mao's death, when Ms. Jiang was almost 30 years old, she was allowed to leave China and come to the United States to study. It was Ms. Jiang's promise to herself that she would share her story with others, made when she was young, that led her to write this book. Ms. Jiang was greatly influenced by *Anne Frank: Diary of a Young Girl* in deciding to tell her story from the perspective of a child instead of an adult looking back on her past. Ji-Li Jiang has said that she hopes the experience of writing this novel and remembering her past will help her to see the future more clearly. Ms. Jiang currently lives in the San Francisco Bay area of California.

Beyond the Book...

CULTURAL REVOLUTION: Read more about China during Mao's Cultural Revolution. Then read up on U. S. history during the same time—1966 to 1976. Make a time line to compare events and movements in the two countries. You might also read more about the history of China and the life of Mao Zedong.

GLOSSARY: At the end of *Red Scarf Girl,* the author includes a glossary of terms the reader should know to understand the book. If you were writing an autobiography to be read by people from other cultures, what words might you include in your glossary? On your own or with other group members, create the glossary you might include at the end of your book. It will be interesting to compare and contrast your glossaries and discuss your reasons for including specific terms.

REFRESHMENTS OR FOOD MENTIONED IN THE BOOK: For lunch or dinner, serve sweet green bean soup, or shrimp with rice and vegetables. For dessert, have pomegranates and popsicles.

IF YOU LIKED THIS BOOK, TRY...

China and Mao Zedong, by Jack Dunster—Dunster describes Mao's rise to power and his regime during the same time frame in which *Red Scarf Girl* takes place.

Let One Hundred Flowers Bloom, by Jicai Feng, Christopher Smith (translator)—A talented young artist battles the propaganda that has arisen against him in a story about growing up during the Cultural Revolution.

The White-Haired Girl: Bittersweet Adventures of a Little Red Soldier, by Douglas Childers, Jaia Sun-Childers (narrator)—The personal and historic events that shaped Jaia's life and country come alive in this memoir of a girl coming of age during the Chinese Cultural Revolution.

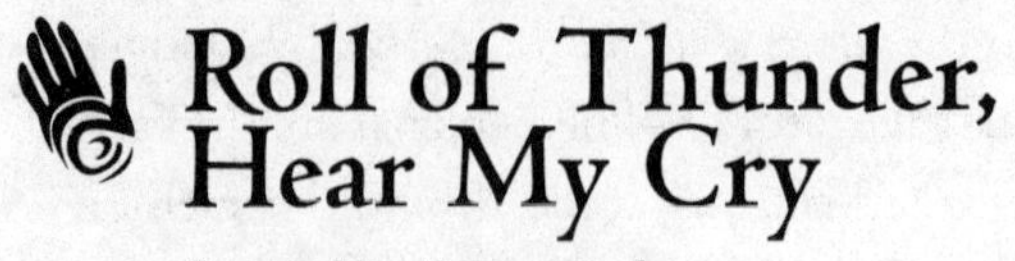

Roll of Thunder, Hear My Cry

by Mildred D. Taylor

Growing up in rural, segregated Mississippi, confronting both poverty and prejudice, a young girl learns how to fight for herself. Despite her struggles, and the daily difficulties her family faces simply to survive, Cassie Logan manages to maintain her dignity and determination in this gripping story.

READING TIME: 2–3 hours, about 276 pages. Some of the subject matter, which includes the Ku Klux Klan, chain gangs, and lynchings, may be too strong for some girls, and mothers may want to read the book first.
THEMES: race, justice and injustice, loss, dignity, self-esteem, family, loyalty

Discussion Questions

✦ How are Cassie and her family's lives touched by the past? Why do you think white people in that time and place are trying to return to pre-Civil War conditions, even though it's the middle of the Great Depression in 1933?

✦ When Cassie and her brothers are teased by the white children riding on a schoolbus, how do they feel? How do you think you would feel if you were in that situation, with dilapidated books and classrooms, when other children had new supplies? Why is it important when Cassie's mother, Mary, changes the front of the books?

✦ How do Cassie and her family feel about having to "know their place"? What do you think it would be like to have to be so careful about everything you said and did? How would that affect your life?

✦ Why is it considered a radical gesture when Cassie's parents encourage their neighbors to boycott the local store and buy their supplies in another town? Would you have done what the Logans did?

✦ Cassie's father tells her, "There are other things, Cassie, that if I'd let be, they'd eat away at me and destroy me in the end. ...There are things you can't back down on. Things you gotta take a stand on." What are those things for Cassie? What would they be for you? Discuss what issues are worth taking a stand on, and why.

✦ Stacey, Cassie's brother, takes a public whipping from his mother rather than betray his friend, TJ. Why does Stacey do that? What would you have done? What kind of friend is TJ to Stacey anyway? Has a friend ever put you in a similar situation?

✦ Why is the land so important to the Logans? Is there something that carries the same significance for your family?

✦ How does visiting Mr. Berry affect Cassie and her brothers?

✦ How does it make Cassie feel to have to wait in Barnett's store until after the white children are helped? If you were treated like that, how would you feel?

✦ Cassie's mother describes Mr. Simms as "one of those people who has to believe that white people are better than black people to make himself feel big." Do you know people who think that way, whether about their religion, ethnic group, or some other distinction? Why do you think some people are like that?

✦ Cassie's father discourages his children from being friendly with Jeremy Simms because he's white. How does racism affect both the white and black people in Cassie's community? Discuss Uncle Hammer's attitudes towards white people, and Mr. Jamison's attitudes towards black people.

✦ The Logan family is compared to a fig tree that "keeps on growing and doing what it gotta do. It don't give up. It give up, it'll die." What does this mean? Discuss specific incidents in the book where the Logans act like that fig tree.

ABOUT THE AUTHOR: (see p. 105)

Beyond the Book...

PRESENTATION: Before the group meets, have someone prepare a presentation on the origins of the Ku Klux Klan, "nightriders," and the Reconstruction period to better understand the context of this novel. Include a discussion about the Southern sharecropping system, and how it was designed to keep people, especially poor black people, in poverty.

CHURCH MUSIC: Going to church was an important activity for African-Americans in the rural South. Get recordings of spirituals, hymns, and gospel music to listen to. If possible, attend services at a local church with an African-American congregation, or arrange to go to a tent revival meeting when one comes to your community.

MOVIE: Rent *Places in the Heart* with Sally Field, which deals with similar issues of poverty and racism in rural Texas.

REFRESHMENTS OR FOOD MENTIONED IN THE BOOK: Something as simple as butter required strenuous activity in the Logan home. Churn your own butter to see how hard it is to make. Pass around licorice, bananas, and oranges, since these were once-a-year treats in the Logan household.

IF YOU LIKED THIS BOOK, TRY...

Song of the Tree, by Mildred Taylor—This is another novel about the Logans.

To Kill a Mockingbird, by Harper Lee—Racism and prejudice affect a Southern white girl during the same era.

Some Other Books by Mildred D. Taylor:

The Friendship

Let the Circle Be Unbroken

Running Out of Time

by Margaret Peterson Haddix

Ten-year-old Jessie Keyser lives in Clifton, IN, in the year 1840—or so she thinks, until she makes the shocking discovery that it's actually 1996, and Clifton is really a historical preserve where tourists can observe the inhabitants through hidden cameras. At first, people come to live there by choice, agreeing to inform children of the reality of their situation only after the age of 12. Medicine and supplies are provided from the "outside" when needed. Now, when a diphtheria epidemic breaks out, exits are guarded and the necessary medicine is not available. Jessie escapes to the present, and makes the even-more-shocking discovery that Clifton is really a cover for a controversial genetic research project. By getting her story to the press, Jessie saves her friends and neighbors.

This is what I call a mind-expanding book. It makes you question all you know to be real, something that is scary and fun at the same time. It is sort of like the game I sometimes play with Morgan and her close friend Brittany, who are so alike it borders on uncanny. Brittany's mother and I sometimes tease the girls that they are really long-lost sisters who were split up and adopted by us at birth. We all laugh for a little while, but the girls always stop the game before long because it's too unsettling to question their reality—even in jest.

READING TIME: 2–3 hours, about 185 pages
THEMES: medical ethics, family, trust

Discussion Questions

- Jessie likes being the only child in the family who helps her mother with her nursing work, even though she is not the oldest. Which personality traits of Jessie's make her mother place so much confidence in her?

- ✦ How does Jessie react when her mother tells her the truth about Clifton? Put yourself in Jessie's place. How does hearing the truth make you feel about your parents? Is your trust in them shaken, or can you accept their justification? If you were a parent, would you have made different choices? For what reasons?
- ✦ When Jessie escapes from Frank Lyle's apartment, at first she thinks she won't be able to climb down the wall, but she succeeds in doing so. Have you ever been in a situation where you thought you would fail, but were able to succeed? What helped you to do something that you thought was impossible? Did your experience have an effect on your self-confidence or self-esteem afterward? Talk about what happened.
- ✦ Frank Lyle (a.k.a. Isaac Neeley) justifies sacrificing the lives of some Clifton residents by saying that the Clifton project, in the long run, will save lives by strengthening the human gene pool. Do you agree that sacrificing the lives of a few individuals in order to save large numbers of people is justifiable? Give reasons for your opinions.

ABOUT THE AUTHOR: Margaret Peterson Haddix was born in 1964 in Ohio, and grew up in a house full of storytelling. Her father's tales of his own childhood encouraged Ms. Haddix to tell her own stories, but she took the process a step further by writing it all down. A summa cum laude graduate of Miami University, she became a journalist after college. *Running Out of Time* was inspired by one of her newspaper accounts of a restored historical village. The idea for her second book, *Don't You Dare Read This, Mrs. Dunphrey,* came from a newspaper series about abused children. Haddix has also written *Leaving Fishers* and *Among the Hidden.*

Beyond the Book...

HOMETOWN HISTORY: Find out about the history of your city or town. If your hometown were to be restored to the way it was 150 years ago, what would one of the main streets look like? Describe a historical restoration of your town in words or pictures.

TIME TRAVEL: As you go through one day, imagine that you are a time traveler from 100 years ago. Which objects would be unfamiliar to you? Which concepts would you have difficulty understanding? Which customs or types of behavior would surprise or shock you? Discuss your day as a time traveler from the past.

MEDICAL ETHICS: Recently, with advances in science and medicine, medical ethics has become more controversial than in the past. For example, some scientists support the cloning of human beings, while others think it should be against the law. Discuss your opinions about cloning and other controversial scientific and medical developments.

REFRESHMENTS OR FOOD MENTIONED IN THE BOOK: Try some beef jerky, bread, and anise cookies.

IF YOU LIKED THIS BOOK, TRY...

The Cellar, by Ken Radford—Sian travels back in time to solve a mystery in an isolated house in North Wales.

Mazemaker, by Catherine Dexter—Twelve-year-old Winnie finds a maze that leads her back into the past.

Some Other Books by Margaret Peterson Haddix:

Don't You Dare Read This, Mrs. Dunphrey

Leaving Fishers

The School Mouse

by Dick King-Smith

A family of mice living in a schoolhouse have a rodent scholar in their midst—their beloved Flora, who figures out how to read right along with the human children. None of the other mice see much use for reading in their little mouse lives until Flora reads the word "poison" on a bag and saves some of her family.

READING TIME: 2 hours, about 123 pages
THEMES: school, literacy, family

Discussion Questions

- What does it mean that Flora's favorite word is "why"? How does this curiosity affect her? What would you say your favorite word is?
- What incites Flora to start paying attention to words and letters? How does she finally make the breakthrough from seeing just little black marks to seeing words and letters?
- Flora's reading ability saves her life and the lives of some of her family. What situations are there in human life where reading can be just as important?
- Besides helping her to save her family from the poison, Flora gets enjoyment out of reading. Discuss the experience of reading for fun. What about reading do people find pleasurable? What is the use of reading? What would life be like if people couldn't read?
- Were you taught to read by a family member or have you taught a family member to read? Discuss these early reading experiences.
- *The School Mouse* is an irresistible, charming book that readers of any age will enjoy. What are its special qualities?

ABOUT THE AUTHOR: Dick King-Smith had two other careers before he became a writer: teacher and farmer. Although he came to writing later in life, his writing career has been very successful. He has written many books for young adults which have earned numerous awards and accolades. Known for creating animal characters with unique personalities and depth, Mr. King-Smith's books have become very popular. Mr. King-Smith still lives in Gloucestershire, England, where he was born and raised.

Beyond the Book…

PET MICE: Invite pet mouse owners to bring their pets to enliven the book discussion. Ask the pet owner to discuss how her mouse is similar or dissimilar to Buck, the pet mouse in the book. You might even want to research the way wild mice actually live.

LEARNING TO READ: The description of Flora learning to read is very much the story of any child learning to read. Share memories of the first time you "cracked the code" of those little black marks on the page—letters, words, and sentences. If you have helped teach youngsters to read, discuss what the process involves.

LITERACY: Volunteer with a local literacy organization, continuing education program, or library in their adult literacy program. After discussion meetings, you may also want to donate books to schools, libraries, or literacy-advocacy groups.

REFRESHMENTS OR FOOD MENTIONED IN THE BOOK: You might enjoy sharing typical school snacks or school lunches in honor of the school mice who feasted on the foods that the children left behind at the end of the day.

IF YOU LIKED THIS BOOK, TRY…

I Can Read with My Eyes Shut, by Dr. Seuss—This timeless book about the joys of reading makes an enjoyable read-aloud after the book discussion of *The School Mouse.*

The Mousewife, by Rumer Godden—This delightful book portrays the domestic life of a mouse family.

Pigs Might Fly, by Dick King-Smith—A pig's ability to swim saves the day.

Stuart Little, by E. B. White—A mouse is born into a human family.

The Cuckoo's Child, by Suzanne Freeman (see p. 55).

Some Other Books by Dick King-Smith:

Babe: The Gallant Pig

The Secret of Gumbo Grove

by Eleanora E. Tate

Raisin, an inquisitive African-American girl, befriends an older woman despite objections from her family and friends. This older woman helps her discover important aspects of her own heritage, and in the process, provides her with the knowledge and skills she needs to move ahead towards her own destiny.

READING TIME: 2–3 hours, about 199 pages
THEMES: race, prejudice, friendship, family, self-esteem

Discussion Questions

- ✦ What makes Raisin different from her sisters and friends? Why does she have a special feeling about history that most of the people she knows simply dismiss?
- ✦ What kind of relationship does Raisin have with Miss Effie? How is it important to Raisin? Have you ever known an adult who offered that kind of friendship to you? If so, what was it like and what did it do for you?
- ✦ Miss Effie tells Raisin, "You got to know about the past to prepare for the future." What does she mean? What does Raisin think about this? What do you?
- ✦ In Gumbo Grove, the adults around Raisin have selective memories about the past, and many of her friends are ignorant. How and why does Miss Effie want to change that? Are there family stories and secrets that you've only been able to glimpse in small snatches? Why do you suppose people sometimes keep that past hidden?
- ✦ How does Raisin's research into the old, black cemetery change her sense of herself and her possibilities? Have you ever learned something about your family that affected you in a similar way? If so, share this experience.
- ✦ Raisin, when she's helping her father sell the crabs he's bought, thinks about how Reverend Walker once likened some black peo-

ple to crabs in a barrel: "He said that's how some Black folks were. Didn't want anybody else to move up the ladder and out of the barrel, trying to pull them back down into the barrel with the rest of them." What does he mean? Can you see that in some of the characters in the book? Can you see that in your community?

✦ How do other people's opinions about Raisin affect how her father feels about her work with Miss Effie?

✦ How does Raisin's relentless pursuit of the past change her community's self-image? Why is uncovering the past problematic for so many people? Why should it matter what someone's ancestors did as far as current accomplishments are concerned?

✦ Do you think who your ancestors were makes a difference in how you live your life? In what way? How?

ABOUT THE AUTHOR: Eleanora E. Tate was born in Canton, MO, and her books are influenced by her rural upbringing. The award-winning *The Secret of Gumbo Grove,* is part of a trilogy, including *A Blessing in Disguise* and *Thank You, Dr. Martin Luther King, Jr.* Her first young adult novel, *Just an Overnight Guest,* was adapted into a film version starring Richard Roundtree. Tate lives in Morehead City, NC, with her husband, Zack Hamlett III.

I agree with Miss Effie that "you got to know about the past to prepare for the future." This is what the story is all about.

Beyond the Book...

COMMUNITY SERVICE: See if there is an old cemetery in your community that has fallen into disrepair, and plan to spend some time helping to clean off the grave markers. If possible, contact your local historical society to research the background stories on people who are buried there, and compile your own history.

GRAVE RUBBINGS: Go to a local cemetery with some crayons and paper. Find interesting or beautiful grave stones and make rubbings of their inscriptions.

FAMILY TREE: Make a project of preparing a family tree, complete with as many family stories as possible. Share it with the rest of your family. Illustrate the tree with drawings or photographs of family members.

PAGEANT: Each contestant in the Miss Ebony contest has to learn about part of her heritage and talk about a role model. Develop your own "pageant" for mothers and daughters to participate in, where each pair tells of meaningful role models in their own lives. Discuss what makes someone a role model. If you want, limit your choice to a family member who has a strong influence. Or have the daughters "judge" the mothers on specific skills, for a fun, role-reversing kind of friendly competition.

RESEARCH: Research the beginning of the Civil Rights movement in the South. Talk about what it would have meant for Raisin's ancestors to live under the shadow of the Ku Klux Klan, and how it would have affected their outlook and daily lives.

IF YOU LIKED THIS BOOK, TRY...

Zora Neale Hurston's novellas or short stories—They have as their theme the experience of African-Americans "passing" for whites, like the pivotal episode in *The Secret of Gumbo Grove.*

Thank you, Dr. Martin Luther King Jr., by Eleanora E. Tate—This is a companion novel to *The Secret of Gumbo Grove.*

Some Other Books by Eleanora E. Tate:

A Blessing in Disguise

Just an Overnight Guest

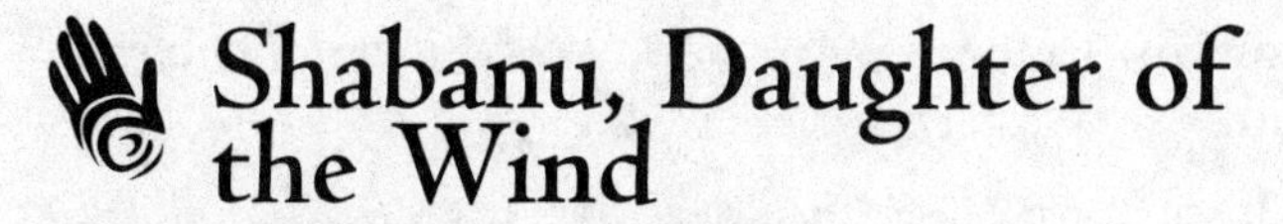

Shabanu, Daughter of the Wind

by Suzanne Fisher Staples

Strong-willed Shabanu is at home in the wind-swept desert of Pakistan where she and her family are nomadic camel herders. Although she is more interested in riding camels than doing housework, she accepts her arranged, future betrothal to Murad, her older sister Phulan's future husband's brother. When Phulan's marriage plans are tragically destroyed, Phulan is married to Murad instead, and Shabanu is betrothed to a wealthy potentate, old enough to be her grandfather.

Like Shabanu, my daughter Morgan dances to the beat of a different drummer. Finding a balance between her strong-willed self and society's expectations can be a struggle in any culture, and girls can learn some important lessons from Shabanu's story.

READING TIME: 4–5 hours, about 240 pages. Some of the themes and content of this book may not be appropriate for younger readers.
THEMES: cultural identity, family, child abuse, coming of age, grief

Discussion Questions

✦ Auntie says to Shabanu's mother, "If God had blessed you with sons, we wouldn't have to break our fingers over wedding dresses." What customs and traditions revealed in the book make sons preferable to daughters in Shabanu's culture?

✦ Do you think Shabanu's father is wrong to sell Guluband, Shabanu's beloved camel, even though he knows how Shabanu feels about him? What does Dadi have to consider besides Shabanu's happiness?

✦ Shabanu is betrothed to a man she does not want to marry. What circumstances make it impossible for Shabanu to assert her own will and simply refuse to marry Rahim-*sahib*? What do you think about the custom of arranged marriages? Would it be possible to have such a custom in Western societies?

- ✦ In several episodes in the book, Shabanu's father beats her. In western culture, how would Dadi be judged as a father? How is Dadi portrayed in the book—as a cruel, brutal child abuser, or as a caring, loving father? What details in the book does the author include to show that Dadi loves Shabanu deeply? How can these seeming opposites—a good man and father who brutally beats his daughter—be reconciled? How do you judge Dadi as a person and as a father?
- ✦ Recall Phulan's wedding and the rituals and ceremonies the author describes. Phulan's wedding may be quite different from those in your experience. What are some of the important differences? Are there any similarities? What are they?
- ✦ Imagine that Shabanu is suddenly transported to a western country, such as England or the United States. What customs and values would she find particularly strange? Would she be happier?
- ✦ Would she have trouble adjusting to customs and values different from her own? Which ones might be a problem for her?
- ✦ What incidents in the book show that Shabanu is growing up and becoming more mature?
- ✦ Discuss how Shabanu's thoughts, words, and feelings allow you to relate to her as a character, even though she comes from a world radically different from yours.

ABOUT THE AUTHOR: Before she became an author, Suzanne Fisher Staples worked as a journalist, a career that took her to Hong Kong, Pakistan, Afghanistan, and India. The time she spent in Pakistan allowed her to come to know the nomads of the Cholistan Desert, the inspiration for this book. Ms. Staples currently lives in New York City.

Beyond the Book. . .

COOKING: Even while setting up a temporary "camp" in the desert, women make *chapatis,* the flat, pancakelike breads cooked on a griddle. Find a recipe for *chapatis* in an Indian cookbook and try making some. While cooking and tasting, think about how your cooking and eating facilities compare with those described in the book.

CHOLISTAN: Find out more about Cholistan from an atlas and an encyclopedia—maybe even a travel agent. Try to find pictures of the desert described by Shabanu in the book. Draw pictures of scenes Shabanu describes, or build a diorama. It might also be interesting to find out more about Islamic religion and culture. For example, why does Dadi say that everything is decided by Allah? Why are women expected to wear *chadors* and cover their faces in the company of men? Why are men permitted to have more than one wife? Why does Shabanu's family fast during *Ramadan*?

WEDDINGS: Do a research project on "Weddings Around the World," and explore weddings in different cultures or locations (including your own). Discuss what you can infer about a culture from its wedding customs. Mothers: bring out your wedding albums and share your memories of that day.

REFRESHMENTS OR FOOD MENTIONED IN THE BOOK: Serve a traditional Indian meal, with spiced lentils, curried chicken, Basmati rice, and vegetables.

IF YOU LIKED THIS BOOK, TRY...

Haveli, by Suzanne Fisher Staples—In this sequel, Shabanu, now married to Rahim, must thread her way through a maze of conflicting loyalties to her husband, her young daughter, her friend, her family—and herself.

Sister of the Bride, by Beverly Cleary—All the excitement and confusion an approaching wedding brings to a household is the basis for the plot of this sensitive book. *Sister of the Bride* provides many contrasts to *Shabanu, Daughter of the Wind.*

Islam, by Christopher Barlow—This is an introduction to Islamic history and the religious beliefs of Islam.

Some Other Books by Suzanne Fisher Staples:

Dangerous Skies

The Shimmershine Queens

by Camille Yarbrough

Angie's father has just moved out and her mom is depressed. In addition, there's a gang at school that taunts her about her dark skin and kinky hair. Daydreams seem the only escape for Angie, until 90-year-old Cousin Seatta comes to visit. Cousin Seatta helps Angie achieve her dreams by showing her how to speak up for herself and discover her own inner strength. Angie and her best friend, Michelle, use what Cousin Seatta calls their "shimmershine feeling" to become leaders and really shine in the class dance production.

READING TIME: 2–3 hours, about 140 pages
THEMES: self-esteem, courage, friendship, racial pride, family, coming of age

Discussion Questions

- What are your words for what Cousin Seatta calls the "get-up gift"? Does everyone have this gift? Cousin Seatta says that we are born with this gift; that it "come down from elders." Where do you think your "get-up gift" comes from? When do you use it?

This book led our club into a pretty deep discussion about self-esteem and friendship.

- What are Angie and Michelle referring to when they talk about the "shimmershine feeling"? Have you ever had the "shimmershine feeling"? If you feel comfortable talking about it, describe the feeling and explain how and when you get it. How has the "shimmershine feeling" helped you to overcome difficulties or achieve difficult goals?

- Cousin Seatta tells Angie a lot about the history of black people in the United States. She says, "We got to remember da old ones, da first ones come off da slave ships from Africa, find out what day know 'bout us dat kept um strong." Do you agree that it is important to know the history of your ancestors? How does this knowledge help people living today?

- Angie finally speaks up to Charlene: "...even if I am scared, if you start sumpum, little scared ugly Angie gonna stop it. Now, make you move." What gives Angie the courage to stand up to Charlene? What makes Charlene back down, even though she is stronger, bigger, and tougher than Angie? How did you feel when you read this passage?

- Reread the passage in which Michelle's mother talks about how black women have tried to straighten their hair, and how their "hair keeps goin home. ...Africa keeps claimin us, hair an all. ..." What does Michelle's mother mean? How does she think African-American girls and women should wear their hair? Why does she think this way?

- What makes Angie take the lipsticks and makeup from Woolworth's? Have you ever had feelings that you didn't know what to do with? Did those feelings ever make you do something you wouldn't normally do?

- After Angie is caught stealing, her mother explains to her that feeling bad about her father is no excuse. She says, "You gon be hurt a lot in life. ...But you can't become a thief because you afraid and confused and hurt. ...That's what growin up is all about!" Do you agree that growing up means holding on to your values even when you're feeling hurt or life is going badly? What else do you think "growin up is all about"?

- What is the most important lesson Angie learns from Cousin Seatta? How does it change her life? What did you learn from reading *The Shimmershine Queens*?

- Do you have a "Cousin Seatta" in your family—an older person who keeps the younger family members in touch with their past, or passes on words of wisdom to the younger generation? What have you learned from your "Cousin Seatta"?

ABOUT THE AUTHOR: Camille Yarbrough, born in 1938 in Chicago, is a dancer, actress, drama teacher, and singer, as well as a prolific writer of books, poetry, and songs. In addition to *The Shimmershine Queens,* Ms. Yarbrough wrote *Cornrows,* a book of children's poetry. In 1972, a program of her poems and songs was presented by Nina Simone at New York's Philharmonic Hall. She is currently Professor of African Dance and Diaspora in the African Studies Department of City College in New York City.

Beyond the Book...

TAPE RECORDING: Ask an older relative or family friend to tell you stories about his or her youth, or about the history of your family. With this person's permission, tape record the interview and play it for the group. Discuss what you can learn from the interview.

SHARING TRADITIONS: As Angie and Michelle learned the Adewa dance, learn a dance, song, tale, custom, or traditional craft that your ancestors knew or practiced long ago. Share what you have learned, and talk about its meaning.

COPING WITH DIVORCE OR SEPARATION: Angie is upset because her father and mother have separated. This happens in many families nowadays. In the library, find a book for children on how to cope when parents divorce or separate. Evaluate the books you have read for how they deal with these topics.

REFRESHMENTS OR FOOD MENTIONED IN THE BOOK: Ms. Collier was teaching the class about Ghana. Continue the lesson by serving traditional West African foods like Jolof rice (similar to pilaf). Or serve one of the girls' favorite treats—ice cream.

IF YOU LIKED THIS BOOK, TRY...

Forever Friends, by Candy Dawson Boyd—Jessie develops self-esteem as she proves to herself and to her father that she can succeed at a performing-arts middle school.

Childtimes: A Three-Generation Memoir, by Eloise Greenfield—These are childhood memoirs of three black women—grandmother, mother, and daughter.

Some Other Books by Camille Yarbrough:

Cornrows

The Little Tree Growin' in the Shade

The Slave Dancer

by Paula Fox

Jesse Bollier often played his fife on the New Orleans docks to earn a few pennies, until one day he was discovered by a sailor and kidnapped. The sailor compelled him to serve on a slaver's ship, playing music for the slaves to "dance" to. This harrowing journey forces him to challenge some of his most basic assumptions about himself and his society, as he learns how to survive by thinking for himself.

READING TIME: 3–4 hours, about 127 pages. Some of the marine and sailing terminology may require additional explanation for clarity.
THEMES: trust, loyalty, friendship, values

Discussion Questions

- Why does Jessie go to Congo Square even though it is forbidden? What does this suggest about his character?
- How does Jessie feel about his mother and sister? How does he feel about his Aunt Agatha? Would you have felt the same way in his circumstances?
- Imagine being kidnapped and taken away from your family, like Jessie is. How would you feel? How does Jessie feel?
- What is Jessie's first reaction to Purvis? How do these feelings change?
- Appearances aren't always what they seem on board *The Moonlight,* where Jessie has to negotiate an ever-shifting web of deceit, lies, and uncertain truths. How does this confusion affect him? What does he learn from the experience?
- Jessie has been brought on board to play his fife so that the slaves can exercise and keep fit during the long sea voyage. How does Jessie feel about his skills being used for this purpose? How would you feel if you were in the same situation?
- Jessie has grown up with certain beliefs about the slave trade. Sailing on board *The Moonlight* challenges those ideas. Specifically,

what is his initial opinion of the slave trade, how does his opinion change, and what specific episodes affect his thinking?

- Why is Nicholas Spark an important influence for Jessie? How does witnessing Spark kill a slave affect Jessie?
- Why is the black slave boy significant to Jessie's development?
- What kind of relationship do Purvis and Captain Cawthorne have? On what are you basing your impressions? How does their relationship affect Jessie?
- Why is the theft of the Captain's egg significant? How is it a turning point for Jessie? How does this change his perceptions of the crew?
- How does Stout's treachery concerning the fife affect Jessie?
- Jessie says that memory and imagination help him escape the horrors of the ship, at least in his mind. Have you ever tried to escape through your imagination?
- How do Jessie's feelings and memories about his mother and sister change while he's on the ship? Why do you think this happens?
- What kind of relationship does Jessie have with Daniel? Why is it important?
- Music has always been important to Jessie. How does his experience on board the slave ship change the way he feels? Have you ever had a passion, like Jessie's for music, that ended up this way? Discuss.

ABOUT THE AUTHOR: Paula Fox was born in New York City in 1923. Her family moved around a lot while she was growing up, which included a two-year stay on a Cuban plantation when she was eight. The young Ms. Fox found solace in the library and in books. When she was young, she was very influenced by stories of her grandmother's life in Spain. Though some of them were humorous, they were all somewhat sorrowful, and seemed to be "mourning for the past." She wanted to write from the time she was very young—but didn't start until she had a job teaching troubled children—and enjoys writing about "children as they encounter the daily surprises of life." Ms. Fox now lives with her family in Brooklyn, NY.

Beyond the Book. . .

MARITIME CENTER: Visit a maritime center, where old sailing ships are on display, to get a sense of the cramped quarters in which sailors and/or slaves lived.

HISTORICAL FICTION: Write a historically based fictional story that takes place on a ship. First choose an event or time you would like to focus on, and research what it was like to live during that period, including the specifics, like what people wore, who the important figures were, what people ate. Most historical fiction uses real people or events even though the story is fictional.

ACTIVITY: Make an oil lamp, like the kinds that were used on slave ships, or try to construct a ship in a bottle from a kit.

MOVIE: Watch the Steven Spielberg movie *Amistad.* For older readers, rent *Mutiny on the Bounty.*

REFRESHMENTS OR FOOD MENTIONED IN THE BOOK: Try some of these sea rations: tea, biscuits, lentils, raisins, and bread.

IF YOU LIKED THIS BOOK, TRY. . .

Rime of the Ancient Mariner, by Samuel Taylor Coleridge—This maritime tale is good to read aloud (for older readers).

Billy Budd, by Herman Melville—This story also deals with a sailing experience.

The True Confessions of Charlotte Doyle, by Avi—A young girl's shipboard experience transforms her life (see p. 266).

To Be a Slave, by Julius Lester

Some Other Books by Paula Fox:

Amzat and His Brothers: Three Italian Folktales

The Eagle Kite

A Likely Place

Something Terrible Happened

by Barbara Ann Porte

Gillian lives in New York City with her West Indian mother and grandmother. Her father, who was white, died when she was a baby, and now Gillian's mother has been diagnosed with AIDS. Gillian is sent to live with her white aunt and uncle in Tennessee. This is the tale of how Gillian wins her fight for emotional survival in the face of tragedy and upheaval.

Our group discussed this novel in February 1998. The girls (and the moms) gained a new insight into homelessness and were forced to give a face to the faceless. The book also sparked a discussion of interracial marriages and multiracial children, and the extent to which race matters in today's world.

READING TIME: 2–2 1/2 hours, about 214 pages
THEMES: AIDS, loss, aging, homelessness, cultural identity, coming of age

Discussion Questions

- Gillian's grandmother says, "I never knew a man to be there when you needed him." Why does she say this? Talk about the men mentioned in the book who fit this description. Then discuss men you know who are around when you need them, including male characters in the book. Extend the discussion and talk about whether or not women *need* to depend on men. How have women's lives been affected by increasing independence in recent history?
- Have you seen or heard about homeless people? How do most people react to them? Now that you have read about how Gillian and her mother became homeless, talk about possible ways other people might lose their homes and be forced to live on the streets. Discuss possible solutions to the problem of homelessness.
- Recall the story of the girl and the carp that Gillian tells to DeeDee. What is the meaning of the story? How does it apply to Gillian's life? Does it apply to your life, and if so, how?

✦ Antoine tells Gillian about her uncle's and father's difficult childhoods. Uncle Henry survived, but Gillian's father did not. Discuss why you believe or don't believe that a person's past determines what her or his life will be like. Is having a difficult past an excuse for destructive or immoral behavior later in life?

✦ Gillian is just reaching puberty—the age at which boys and girls become physically mature. Naturally, she is interested in and concerned about the changes she is beginning to notice in her own body. If you feel comfortable with this topic, talk about the physical and emotional changes that occur in early adolescence.

✦ Death is a difficult topic to talk about, but it is a prominent topic in the book. If it doesn't make anyone uncomfortable, talk about personal experiences with loss, and the feelings and thoughts that may accompany such experiences.

✦ AIDS is another difficult but prominent topic in the book. Discuss what you have learned about AIDS and how to prevent contracting the HIV virus. You might also talk about people who have AIDS or are HIV-positive and the problems they face.

✦ Gillian is proud of her grandmother for becoming both a runner and a Ph.D. candidate. Discuss older people you know—acquaintances or family members. Do any of them contradict the stereotypes many people have of the elderly?

✦ Recall the story of King Solomon and the magic ring that spoke these words, "This too shall pass." Why would this ring make a sad person happy, and a happy person sad? Do you think, "This too shall pass" is a helpful saying to remember throughout life? Why, or why not?

ABOUT THE AUTHOR: Barbara Ann Porte's parents encouraged her and her two sisters to read and tell stories from an early age. Growing up in New York City, Ms. Porte and her sisters helped out in their father's drugstore. She earned a degree in agriculture and later returned to school to become a librarian. Ms. Porte began writing after her youngest child finished college. She has written numerous children's books and currently lives in Arlington, VA.

Beyond the Book...

STORIES: Gillian knows many stories her grandmother or others have told her. Each story teaches a lesson or important idea about life. Read other stories that teach lessons. Aesop's fables and the fables of Lafontaine are good ones to start with. Then make up your own stories. Think about a *moral,* or lesson, you would like your story to teach. Then choose either human or animal characters and make up the plot. Gillian enjoys telling stories to her younger cousin DeeDee. You can extend this activity by telling stories you've read or made up, to others, perhaps a younger friend or family member.

ART: One of the ways Gillian expresses herself is by making collages. Gillian's collages include photographs, sketches, and print material from newspapers and magazines, as well as other materials. Gillian cuts up her photos and sketches and combines and arranges them in different ways before pasting them down. Try your hand at making a collage, using some of Gillian's techniques if you wish; or you may prefer to develop your own techniques. Experiment with your materials until you feel that your collage expresses something about yourself—ideas, feelings, personal identity, or the like.

IF YOU LIKED THIS BOOK, TRY...

The Kidnapping of Aunt Elizabeth, by Barbara Ann Porte—Ashley has to write her family's history for a school project, in this humorous book.

Up a Road Slowly, by Irene Hunt—After Julie's mother dies, Julie must make the difficult adjustment to living with her aunt in this Newbery Medal–winning novel.

You Shouldn't Have to Say Goodbye, by Patricia Hermes—Sarah's mother is dying of cancer.

Some Other Books by Barbara Ann Porte:

Fat Fanny, Beanpole Bertha, and the Boys

Somewhere in the Darkness

by Walter Dean Myers

Jimmy is 14 years old; his mother is dead, and his father is in prison for murder. But Jimmy is doing okay, living with his grandmother, Mama Jean, in Harlem. Things get complicated, however, when Jimmy's father, Crab, shows up. Crab and Jimmy set off on a trip halfway across the country to find a former friend of Crab's who can tell Jimmy the truth—that Crab never killed anyone. After Crab dies from a chronic illness in an Arkansas hospital, Jimmy returns to New York, having learned a great deal about understanding and forgiveness.

I liked this book because, while Crab didn't have much to give, he wanted to give Jimmy all he had. He realized that though he couldn't fix the past, he may be able to help his son understand, and through that understanding, forgive. It's a heart-wrenching but valuable story.

READING TIME: 3–4 hours, about 168 pages
THEMES: parent-child relationships, understanding, forgiveness, trust, responsibility

Discussion Questions

✦ Does Jimmy want to leave New York with Crab? Why do you think he agrees to go? Do you think the decision to go is the right one? Try to "walk in Jimmy's shoes." Would you have gone with Crab, or insisted on staying with Mama Jean?

✦ What do you think Jimmy hoped his father would be like? Compare and contrast the real Crab with Jimmy's dream of what a father should be.

✦ Jimmy says to Crab, "You don't even know how to be a father!" Why does Jimmy say this? What does knowing how to be a father really mean? What does a person have to know and do to be a good father?

- ✦ Jimmy is afraid of Crab until he hears Crab crying in his sleep. Why does hearing Crab cry make Jimmy stop being afraid of him?
- ✦ Do you think it was difficult for Mama Jean to let Jimmy go with Crab?
- ✦ For daughters: Jimmy wishes he were grown up so that he would "have a mind that had answers where his own mind had questions." What kinds of questions does Jimmy have? Do you agree that adults have all the answers when it comes to solving problems, or do adults have questions too? What kinds of questions about life do you think the adults in your family have? What is your biggest concern about life?
- ✦ For mothers: What kinds of questions about life do you think trouble kids? Which questions troubled you when you were their age? What troubles you the most today?
- ✦ What do you think it takes to raise a child that is not your own? What do you think the fact that Mama Jean took Jimmy says about her relationship with his parents?
- ✦ What do you think of the way this book ended? Would you call the ending happy or sad—or both? Why?

ABOUT THE AUTHOR: (see p. 102)

Beyond the Book...

MAP: On a map of the United States, trace Jimmy and Crab's journey to Chicago, then on to Arkansas, and finally Jimmy's trip back to New York. Using small Post-its™, pinpoint where the major events of the plot occurred. Going back to the text for information, include dates (day 1, day 2, and so on).

JAZZ AND BLUES: Crab tells Jimmy that he plays the saxophone, but he doesn't play jazz or blues well enough to sit in with his musician friends in Chicago. Listen to some contemporary jazz and blues music to get an idea of the kind of music Crab likes. Do you like this music? What kinds of moods do different selections create? Here are some suggestions for jazz: Miles Davis, Dizzy Gillespie, Charlie Parker, Lester

Young; suggestions for blues: Muddy Waters, Bessie Smith, Howlin' Wolf. All of these should be easy to find at your local library.

ROLE-PLAYING: Role-play the scene that might have occurred between Jimmy and Mama Jean when Jimmy returned home to New York. Trying to remain true to character, think about what each one would have said to the other.

IF YOU LIKED THIS BOOK, TRY...

Slam!, by Walter Dean Meyers—In another book by the same author, 16-year-old "Slam" Harris is counting on his basketball talents to get him out of the inner city and give him a chance to succeed in life. But he finds that there's more to life than scoring on the court. Realistic, coming-of-age fiction.

Fly Like an Eagle, by Barbara Beasley Murphy—On a cross-country trip, Ace Hobart develops a deeper and more honest relationship with his father, who is searching for information about his own parents and their reasons for placing him in an orphanage.

Some Other Books by Walter Dean Myers:

Darnell Rock Reporting
Fallen Angels
Fast Sam, Cool Clyde and Stuff

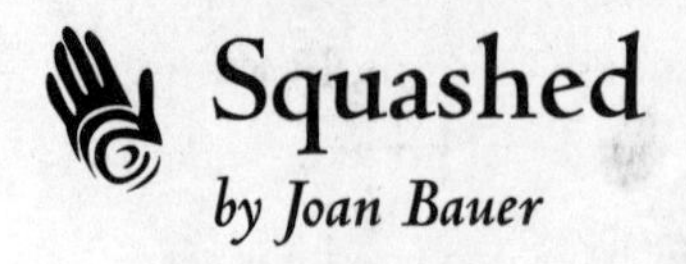

Squashed

by Joan Bauer

Sixteen-year-old Ellie Morgan hides behind her extra weight. She finds satisfaction in her devotion to growing a giant pumpkin, much to the dismay of her father. Ellie's efforts to grow a prize-winning pumpkin give her much-needed focus and are a big boost to her self-esteem.

This sweetly comic and surprising novel will spur conversations about weight consciousness and self-image. The reader will empathize with Ellie, whose perseverance is heartening.

READING TIME: 2–3 hours, about 194 pages
THEMES: acceptance, identity, popularity, jealousy, loss, self-esteem, relationships

Discussion Questions

- ✦ How is Ellie's identity wrapped up with the growth of her pumpkin, Max? Why does she care so much about growing the biggest pumpkin?
- ✦ What kind of relationship does Ellie have with her father? How does he feel about growing things? How does Ellie feel about it? How do their differences create conflict?
- ✦ What does respect mean to Ellie? What does it mean to you? How can you tell that your parents respect you?
- ✦ Ellie observes: "I think what bugs Dad most about me is that I love something he's always resented." What is she talking about?
- ✦ Why is Ellie's relationship with her grandmother so important? What does Nana offer Ellie that no one else can or will?
- ✦ Why does popularity matter to Ellie's father? How does she feel about it? How do you define popularity? Does it matter to you?

- How does Ellie's interest in the new boy in town, Wes, change her behavior and attitude?
- What does Ellie miss most about not having a mother? What kinds of confidences do you and your mother share? What would you miss most if your mother were no longer around?
- How do gardening and growing things connect Ellie to her mother? Are there activities or hobbies you pursue that link you to a loved one who is no longer around? Discuss what those activities might be.
- Why does Ellie's father change his mind about the pumpkin? How does that affect Ellie?
- Ellie's Nana tells her, "Winning's a fine thing, Ellie, but it's all the months and years after that make you who you are." Explain what she means.

ABOUT THE AUTHOR: Joan Bauer was born in River Forest, IL. She began her career as a freelance writer, working in many different fields, including sales, advertising, television, radio, and film. She then moved on to writing nonfiction and screenplays; *Squashed* is her first novel. Ms. Bauer now lives with her husband, daughter, and pets in Darien, CT.

Beyond the Book...

GARDENING: Gardening and growing things are meaningful activities for Ellie in this book. While reading the book, try growing a dish garden of plants or flowers. Coleus (a leafy plant) and beans are fast-growing choices.

PRESENTATIONS: Ellie's father makes his living as a motivational speaker and consultant. As an activity, have the mothers and daughters, in separate groups, discuss what kinds of advice or slogans work for them, and make a presentation during the meeting.

REFRESHMENTS OR FOOD MENTIONED IN THE BOOK: In the book, Ellie takes great pride and delight in her cooking skills. To capture the feeling of abundance that is conveyed by her ambitious menus, serve a variety of desserts—butterscotch ice cream, donuts, cake—at your discussion session.

IF YOU LIKED THIS BOOK, TRY...

The Best Little Girl in the World, by Steven Levenkron—This story for older readers deals with self-esteem and anorexia nervosa.

The Secret Garden, by Frances Hodgson Burnett—A young girl's dedication to gardens helps effect a transformation in her own personality.

Some Other Books by Joan Bauer:

Sticks

Thwonk

The Star Fisher

by Laurence Yep

As a Chinese-American girl living in Clarksburg, WV, during the 1920s, Joan Lee has to learn how to handle prejudice and ignorance, and find acceptance for herself and her family. As she negotiates the often treacherous obstacles placed in her way, Joan develops self-confidence and an appreciation for friendship.

READING TIME: 2–3 hours, about 148 pages
THEMES: discrimination, loyalty, family, self-esteem, identity, popularity, loneliness

Discussion Questions

- How is Joan's parents' outlook on life different than hers? How does that difference affect Joan? Do you ever feel like your outlook on life is different from your parents'? In what way?
- When they first arrive in Clarksburg, Joan feels that she and her family are "monkeys in the zoo." Why does she feel this way? Have you ever been in a situation where you felt like this?
- Some of their new neighbors paint rude words on the Lees' fence. Why do you think they did this? How would you feel if someone did it to your family?
- As immigrants, Joan's parents are less competent in America than Joan and her siblings are. How is Joan's relationship with them affected by their dependence on her to manage certain aspects of their life? What does it mean for immigrant parents to relinquish some of their authority to their children in this way?
- There are times when Joan seems embarrassed and/or ashamed of her mother. What conflict do these feelings raise within Joan?
- At several different points in the book, Joan gains new insight into her mother. What are some examples of when this happens? Have you ever gained new insight into your mother? What was it and how did it come about?

- ✦ Her father is always saying that "the nail that sticks out gets hammered." What do you think he means? What does this imply about how Joan is supposed to behave?
- ✦ For Joan, what does it mean to "fit in"? Why does she care about it? Do you care about fitting in?
- ✦ Joan isn't the only one to feel left out and shunned from the tight social cliques at her high school. How does the other girls' treatment of Bernice, who was the first one to make friendly overtures to Joan, change how Joan feels about their friendship?
- ✦ Why do Bernice and Joan misunderstand each other? What do they have in common? What's different about their situations?
- ✦ Why do Miss Lucy and Joan become friends? What kind of relationship do they have? Have you ever had that kind of friendship with an adult who wasn't part of your family?
- ✦ How does the pie social change the way the town feels about the Lees? Does it change the way the Joan feels about the town?
- ✦ The story of the star fisher is an important one for Joan. Why does Joan identify so strongly with this legend?
- ✦ In the end, who is being excluded from the high school clique? Do you think there was any way for the all the girls to have been friends?

ABOUT THE AUTHOR: (see p. 46)

Beyond the Book...

CHINATOWN: Locate material on the Chinese-American migration to America, which, as this story points out, was not limited to the two coasts' Chinatowns. Find out what other kinds of occupations and businesses Chinese immigrants participated in, besides running laundries.

COOKING EXPERIMENT: Joan's mother wasn't a very good cook. Have the group share stories of cooking experiments that went awry, and how they handled them. At one meeting, members of the group could bake an apple pie together, the way Miss Lucy taught Joan's mother too.

COMING TO AMERICA: Most Americans have an immigrant story in their family history. As part of your discussion, share your families stories. Discuss what motivated your ancestors to come to America, and what kinds of obstacles they faced at first. If there are special family mementos or heirlooms that are significant, share them.

REFRESHMENTS OR FOOD MENTIONED IN THE BOOK: Miss Lucy's tea party causes some consternation for Joan and her younger sister, Emily, who are used to drinking tea Chinese-style. Have an old-fashioned mother-daughter tea party, complete with a fancy tea service, fine china plates and cups, and delicate treats like sugar cookies.

IF YOU LIKED THIS BOOK, TRY. . .

Child of the Owl, by Laurence Yep—This also deals with a young girl's coming to terms with her Chinese heritage (see p. 44).

The Secret of Roan Inish, by Mick Lally—This Celtic variation on the star fisher story is about a magical creature who never quite fits into the human world.

Some Other Books by Laurence Yep:

The Imp That Ate My Homework

Later, Gator

The Last Garden

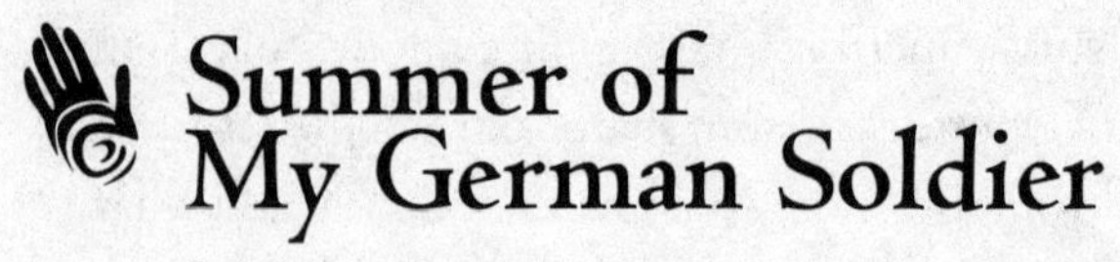

Summer of My German Soldier

by Bette Greene

A 12-year-old girl is forced to confront both personal and political demons in her life when German prisoners-of-war arrive in her Arkansas community. Finally, she must make difficult choices about what she values, and how to reconcile her own beliefs with her community's.

I really enjoyed this book and look forward to discussing it with our club. It covers so many issues, from self-esteem to patriotism, it'll be hard to know where to start!

READING TIME: 2–3 hours, about 199 pages
THEMES: patriotism, obedience, prejudice, individual versus community values, self-esteem, friendship, courage, secrecy

Discussion Questions

✦ How do Patty's parents treat her? How does Patty feel about her parents as a result?

✦ Patty observes that "If there were no mirrors or mothers I probably would never know how ugly I am." How does her mother tell Patty that she's ugly? How would you feel if your mother never seemed satisfied with you?

✦ Appearances matter a great deal in Patty's family. Her mother compares her unfavorably to her younger sister, and her father sees his despised mother in Patty's mannerisms and looks. What do you think about this emphasis on appearance? How does it affect their relationships with each other?

✦ Pearl's mother says to her, "You, Pearl, never liked anything once it was yours." What does that mean? What does Pearl like? How is that true in her relationship with Patty? How does Patty feel about not being "what [her mother] wanted"?

- ✦ Patty feels at "home" only at her grandparents'. Why? What makes it "home" to her? What does "home" mean to you? Do you have a place that seems more like home than your own home? How come?

- ✦ Why does Patty become friends with Anton? How does Patty convince herself and others that it's all right for her to be friends with him, even though she's Jewish and he's a Nazi soldier? How does Patty change because of her friendship with Anton?

- ✦ Why is Patty's relationship with her housekeeper, Ruth, so important to her? What does she get from being with Ruth that she doesn't get from her parents? What are some of the values that Ruth communicates to her?

- ✦ Patty remembers her previous year's Father's Day gift like this: "I knew that he would be pleased with my gift. He'd say it was the finest shirt he'd ever owned. And then the focus would shift from gift to giver and I would rest there in his arms like a long-lost daughter come home." How did he actually respond? How did that affect Patty?

- ✦ Why do people who aren't Patty's parents react to her differently? What does she get from these strangers—a housekeeper and an escaped soldier—that she doesn't get from her family?

- ✦ How do Ruth and Anton make Patty see her world differently? How does Anton change the way Patty sees herself? Have you ever known someone who changed the way you look at things?

- ✦ Patty experiences many conflicts as a result of hiding Anton. She says about her father and Anton, "If I ever had to sacrifice one for the other which one would it be? The one who had fed and sheltered me, or the one whom I had fed and sheltered?" Why is that a hard choice for her to make? What does each of them offer her?

- ✦ What's important about the ring that Anton gives to Patty? How does it make her feel about herself? Have you ever received a present that carried with it this kind of significance? If you did, what made that gift so special?

- ✦ Why does it matter that even after Patty's parents fired Ruth, she is Patty's only visitor while at reform school? How does that make her feel about Ruth? About her family?

✦ How does Patty's nature lead to her downfall?

ABOUT THE AUTHOR: Bette Greene was born in 1934 in a small, rural town in the Arkansas Delta. Growing up Jewish in a poor, Southern town during the Great Depression and World War II was a lesson in difference for her. She now writes about children struggling to maintain their differences against difficult odds. She married Donald Summer Greene in 1959, and has two children, Carla and Joshua. Ms. Greene and her husband now live in Brookline, MA.

Beyond the Book...

WORLD WAR II: Before the group meets, have someone prepare materials on Hitler's Youth groups, and some explanation of Nazi politics and philosophy. Bring in books, maps, and contemporary magazines or newspapers from the time period that discuss World War II and help explain how Americans responded to the war (if they are easily available). Talk about what it might mean if your country were at war, and people thought of as your enemy suddenly moved into your community.

MOVIE: Rent the movie version of *Anne Frank: Diary of a Young Girl.* How was seeing a visual portrayal of her experiences different than reading a literal one?

REFRESHMENTS OR FOOD MENTIONED IN THE BOOK: Food, who prepares it, where it's eaten, and so forth, is important to Patty. She has to draw upon her resources and ingenuity, especially, to get food to Anton. Talk about what it would be like to hide someone in your home, without your parents knowing. What foods would you select? How would you manage to succeed? Serve iced tea and Jell-o™ as refreshments for a meeting.

IF YOU LIKED THIS BOOK, TRY...

Morning Is a Long Time Coming, by Bette Greene—This is the sequel to *Summer of My German Soldier.*

Jacob Have I Loved, by Katherine Paterson—This book deals with another difficult mother-daughter relationship (see p. 153).

Some Other Books by Bette Greene:

Philip Hall Likes Me. I Reckon Maybe.

The Drowning of Stephan Jones

Sun & Spoon

by Kevin Henkes

Ten-year-old Frederick Gilmore, fondly known to everyone as Spoon, secretly continues to grieve over the death of his beloved grandmother, Martha, while everyone else in the family seems to have moved on. So, afraid that he will lose his grandmother again if her forgets her, Spoon sets out to find an object that will keep her memory alive always.

This book reminded me of my own grandparents and the deep and lasting sense of loss I felt when my grandfather, with whom I had a very special relationship, died. Kevin Henkes does a wonderful job of portraying the unique nature of the grandparent-grandchild relationship, and the pain of loss.

READING TIME: 2 hours, about 135 pages
THEMES: death, grief, sibling rivalry, family, coming of age

Discussion Questions

- Why is it so important for Spoon to find something to remember his grandmother by? What keepsakes help you remember family members or friends? Explain why such objects mean so much?
- Why doesn't Spoon tell anyone what he is doing?
- Spoon spends a long time searching for the perfect keepsake of his grandmother. At one point the author says: "And [Spoon] didn't want the 'something' to be a girl thing like a necklace or a pin or an earring." What do you think of this "girl thing" comment? If there are "girl things" and "boy things" that someone might pick as a memento of a loved one, what do you think those objects might be? Do you have any special objects that especially symbolize you? If so, why does this object mean so much?
- Why does Spoon begin to feel that he is losing his grandfather, Pa, as well? Have you had the experience of an older relative losing interest in things they used to do with you?

- ✦ Pa and Spoon have an important desire in common. Both of them desperately want a sign from Gram that she is watching over them. What sign does Pa seem to get from Gram? What sign does Spoon get that his beloved grandmother is communicating with him? Discuss whether you believe in "signs."
- ✦ Each grandchild reacted differently to their grandmother's death. Describe those differences. Do you think Spoon is generally more sensitive than the others? How does Spoon feel about his brother and sister?
- ✦ Here's how the author describes Spoon's sadness after he takes his grandmother's playing cards: "This sadness was overwhelming and specific, and unlike his sadness over Gram's death, was caused by his own actions." Why does Spoon feel this way? Talk about an experience where your own actions caused heartache.

ABOUT THE AUTHOR: While growing up in Racine, WI, Kevin Henkes says he often wondered what it would be like to be an author, and what their daily lives might be like, but never imagined he would become one. When he was 19 years old, his love of books and illustrations led him to New York City to pursue a career as an author. Since then he has written and illustrated many picture books and written numerous novels.

Beyond the Book...

GRANDPARENTS: Invite as many grandparents as possible to the book discussion of *Sun & Spoon.* Talk about the kinds of activities you and your grandparents enjoy doing together. If your relatives can't come or have died, share memories of being with them.

SHOW AND TELL: Extend the book discussion of *Sun & Spoon* by setting aside a Show and Tell time. Bring a memento of an older family member to share, and talk about why the keepsake represents that person.

FAMILY PHOTOS: Search through old family photos for pictures of ancestors taken at approximately the same ages you are now. Point out how Spoon felt the most connected to his grandmother when he really studied a picture of her as a ten-year-old child.

REFRESHMENTS OR FOOD MENTIONED IN THE BOOK: Peanut butter, banana, and cinnamon sugar sandwiches and "Kitty Milk" (honey and chocolate milk) are Spoon's favorite foods at his grandparents' house. Serve these, or everyone's favorite family treats, as you discuss the book.

IF YOU LIKED THIS BOOK, TRY...

Dicey's Song, by Cynthia Voigt—This Newbery Medal–winning novel tells the story of a young girl living with her grandmother.

The Hundred Penny Box, by Sharon Bell Mathis—A fragile, elderly aunt tells about each year of her life as she goes through the pennies in her keepsake box.

Some Other Books by Kevin Henkes:

Bailey Goes Camping

Chester's Way

Taste of Salt: A Story of Modern Haiti

by Frances Temple

Haitian life is harsh under the political dictatorship, particularly for a poor young boy whose efforts to survive are thwarted by obstacles. Djo's attempts to make something of his life offer a heartbreaking glimpse into the realities of poverty and repression.

Taste of Salt is beautifully told through the eyes of two Haitian teenagers. It offers an intense and important look into what is happening in a part of the world most of us are not exposed to.

READING TIME: 1–2 hours, about 179 pages
THEMES: poverty, prejudice, loyalty, faith, family, friendship, loss, responsibility

Discussion Questions

- Witnessing events and telling stories are thought of as powerful ways to change the world in this novel. Are there any stories or experiences you've had that you feel had a similar impact on your life? If so, explain what they were and why they had such a profound influence.
- How are Djo's and Jeremie's experiences different? How are they the same? Why is their relationship so important to each of them?
- Djo's mother, who has many children to feed and little money to do so, sends him out to work so that he can fend for himself. What do you think it would be like to have to take care of yourself, at his age, without your parents taking care of you? How would you develop differently? How would it affect you?
- Djo becomes a follower of the charismatic Haitian priest, Father Aristide (known familiarly as Titid), who ultimately begins to transform the country. The priest is one of the most important influences on Djo, who observes: "Titid says that it is not only the

body, with its feet and hands and strong back that can be useful. He says the mind and spirit are useful, too." What do you think Titid means by that? What does it mean to Djo?

- ✦ What does freedom mean to Djo and the other young boys who live with Titid? What does freedom mean to Jeremie? What does it mean to you?
- ✦ How does Djo feel when he is captured to cut cane in the Dominican Republic? How does his experience there, in veritable slavery, change Djo?
- ✦ What kind of relationship does Djo have with Donay? Why is Donay so important to Djo?
- ✦ What does it mean to take a stand, the way Jeremie does, at great personal risk? Have you ever done anything like that? What motivated you, and what happened as a result?
- ✦ Jeremie, as she sits by Djo's hospital bed after he has been beaten by the police, remembers when Titid preached, "You can only love somebody if you are willing to see them truthfully." What does that mean? Can you see that in your own life?

ABOUT THE AUTHOR: Frances Temple was born in 1945 in Washington, DC, attended Wellesley College from 1963–1965, and served in the Peace Corps from 1965–1967. Her family's involvement in the Sanctuary Movement was the inspiration for her novel *Grab Hands and Run.* Her works include *Tiger Soup: An Anansi Story from Jamaica; Tonight, by Sea; The Ramsay Scallop;* and the award-winning *Taste of Salt.* Ms. Temple died of a heart attack in 1995.

Beyond the Book...

HAITI: Using newspaper and magazine articles from the library or the internet, gather information about what life was like for ordinary Haitians under the dictatorship. Extend the discussion to include other repressive regimes, and how poverty in those countries affects children.

ISLAND MUSIC: Caribbean music and dancing play an important part in the book, as a release from the pressures and routines of daily life. It is one of Djo's talents and helps him when he is imprisoned as a sugarcane cutter. During your discussion, play CDs of popular Haitian music, and during a break, try to learn some of the dances that are mentioned in the novel.

REFRESHMENTS OR FOOD MENTIONED IN THE BOOK: The simplest meals are often beyond reach for Djo and his comrades. Serve rice and beans, and talk about how this ordinary meal would have required such hard work and effort for people in Haiti.

IF YOU LIKED THIS BOOK, TRY...

The Slave Dancer, by Paula Fox—This story offers another look at what it means for a young boy to be captured against his will, and ultimately achieve his own identity (see p. 229).

Some Other Books by Frances Temple:

The Beduins' Gazelle

Grab Hands and Run

There's an Owl in the Shower

by Jean Craighead George

This novel is the story of Borden Watson, whose father has been out of a logging job ever since the government put an end to logging near the Watsons' Oregon home. From Borden's point of view, there's one reason only for his dad's joblessness—the endangered spotted owl which the government is trying to protect at the expense of loggers' jobs. Borden hates the spotted owl for taking away his father's livelihood, until the day he finds a lost owl in the forest and brings it home.

READING TIME: 2 hours, about 134 pages
THEMES: environment, endangered species, ecology

Discussion Questions

- When the book begins, Borden Watson is determined to hunt down a rare and endangered bird, the spotted owl. Discuss your feelings about this situation. Since the issue of Borden's jobless father isn't explained until later, how do you feel about what Borden is doing when the book opens?
- To Borden Watson and his father, there is one reason only that Mr. Watson is out of work: the spotted owl. What do you think of the argument the pro-owl people are having with the loggers?
- Judge Kramer, who has made the decision to end logging, says to Borden: "The owl is telling the people we aren't managing our forest right. Save the owl, and you save the people too." How is it possible that an endangered bird like the spotted owl can save people?
- The author provides many facts about the endangered spotted owl. What did you learn about this bird from the book?
- Borden learns that every living creature and plant connects to every other creature and plant in the forest. Describe how the life of the trees in the forest and the life of the spotted owl are dependent on each other.

- ✦ If the spotted owl becomes extinct, how will this affect the ecology of the forest?
- ✦ Describe Borden's relationship with his father. How does their relationship change after they adopt Bardy, the owl?
- ✦ Have you ever taken in an orphaned or wounded wild animal? Describe your experience.
- ✦ What did you think of the double ending of the book—Bardy is saved, but Enrique gets no answer when he calls for his mate? What do you think the author is trying to say by giving the book these two endings?

This earnest but lighthearted look at the struggle between man and the wild is great for any nature lover.

ABOUT THE AUTHOR: (see p. 158)

Beyond the Book...

VOLUNTEER: As part of the book group or in mother-daughter pairs, encourage members to get involved in saving an endangered species or habitat by fund-raising, working with wildlife groups, or donating money or time. Ask mothers and daughters to think about ways caring individuals can make a difference in improving the environment.

WILDLIFE TRIP: Visit a zoo or get involved in local wildlife groups to go on outings to learn about local natural habitats.

REFRESHMENTS OR FOOD MENTIONED IN THE BOOK: The kids in this book went to a pizza parlor to think. Serve pizza as you're discussing this book.

IF YOU LIKED THIS BOOK, TRY...

All Wild Creatures Welcome: The Story of a Wildlife Rehabilitation Center, by Patricia Curtis—This book is about a wild-animal rescue center for injured or abandoned animals.

The Cry of the Crow, by Jean Craighead George—A fictional story about a young girl who tames an abandoned crow.

Saving America's Birds, by Paula Hendrich—This nonfiction book is about endangered American birds and the methods used to save them.

Some Other Books by Jean Craighead George:

Julie of the Wolves (see p. 157)

Julie

Water Sky

Shark Beneath the Reef

To Kill a Mockingbird

by Harper Lee

This Pulitzer Prize-winning novel describes a young girl's youth in pre–Civil Rights era Alabama. Both a story of her coming-of-age experience and of the scandalous, small-town trial in which her father defends a black man accused of raping a white woman, this book is an exploration of family, community, individuality, and human behavior in general.

I read this book many, many years ago and didn't come back to it again until I was reviewing books for this project. As an African-American, parts of the book stirred emotions from deep within me. Even though the book takes place in the 1930s, the issues of prejudice, violence, and hypocrisy are as relevant today as they were back then.

READING TIME: 5–7 hours, about 281 pages. The somewhat complex and difficult narrative may be challenging for younger readers.
THEMES: race, father-daughter relationships, conformity, morality

Discussion Questions

- How does Scout see her father at the beginning of the book? How does that change?
- How does Scout feel about school? Why does she feel this way? Can you identify with her reaction?
- What are Scout's initial impressions of her mysterious neighbor, Boo Radley? What makes her opinions of him change?
- Why are the Ewells important in Scout's life?
- The African-American housekeeper, Calpurnia, plays an important part in Scout's life. How does Scout relate to Calpurnia? What does their relationship suggest about race relations in the South at that time?
- How do the neighbors react to Scout's father, Atticus Finch, taking on the defense of Tom Robinson, the black man accused of rape?

How do those reactions affect Scout? Imagine being placed in a similar situation. How would you feel?

- ✦ Which values does Scout accept from her community? Which does she reject?
- ✦ What is it like for Scout when Calpurnia takes her and Jem to worship at her church? Does it change the way Scout sees things? How?
- ✦ How does Tom Robinson's trial change Scout? What does this experience mean to her?
- ✦ How are Boo Radley's gifts significant to Scout? Have you ever received gifts that affected you in a similar way?
- ✦ How does Tom Robinson's death affect Scout? Why does she feel the way she does?
- ✦ Maudie tells Scout that it is a sin to kill a mockingbird because they don't do anything more than innocently sing. Why is this a significant statement? Why do you think it is the title of the book?
- ✦ What kind of life do you think Scout will lead as an adult? Why?

ABOUT THE AUTHOR: Harper Lee was born in Monroeville, AL, in 1926. In the 1950s, she moved to New York to write but ended up working as an airline reservations clerk, until friends gave her a Christmas gift: enough money to support herself for a year. During that year she finished the manuscript that would become *To Kill a Mockingbird.* The book is based largely on her childhood, though the trial is fictional. Her father, A. C. Lee, was the model for Atticus Finch, and Harper Lee has described this book as a love story between Atticus and his children. *To Kill a Mockingbird* is Lee's first and only work. She still lives in New York, and travels back to her small hometown in Alabama each year to avoid the Northeast winters.

Beyond the Book. . .

SCAVENGER HUNT: Go on a scavenger hunt for items like the ones Scout and Jem find in the book. Make a purple treasure box to keep them in.

Add any keepsakes or little objects you have that are important to you. Talk about why these things are important to you.

PAPIER-MACHE: Make a papier-mache tree and leave small, secret gifts in it for a family member or friend.

SUNSET: Scout and Maudie like to sit and watch the sun go down. Plan to watch the sunset with one of your neighbors or friends. Bring the book and read aloud the passage where they do this for the first time.

COURTROOM: Visit a local courtroom while a trial is going on to learn about the justice system. If you know anyone who is a trial lawyer, go watch them in action to get an idea of how Scout felt when she watched her father try a case.

MOVIE:: Rent the black-and-white version of *To Kill a Mockingbird.* For older children, rent *In the Heat of the Night,* and discuss the similarities and differences between the two films.

REFRESHMENTS OR FOOD MENTIONED IN THE BOOK: Make a meal of fried chicken and biscuits, or ham and crackling bread. Serve fresh, homemade lemonade.

IF YOU LIKED THIS BOOK, TRY...

Roll of Thunder, Hear My Cry; Song of the Trees, by Mildred D. Taylor—These two books are about the African-American experience prior to the Civil Rights movement (see p. 212).

Inherit the Wind, by Jerome Lawrence—Similar themes of community pressure versus the individual's integrity are dealt with in this book.

Tom's Midnight Garden

by Philippa Pearce

Several boring weeks stretch ahead when Tom is sent off, under protest, to stay with his childless aunt and uncle until his brother recovers from the measles. Aunt Gwen and Uncle Alan mean well, but their small apartment in a renovated mansion hardly provides Tom with enough entertainment to pass the days. Then one night Tom hears a mysterious clock chime a 13th hour, and a doorway leads him, not to the expected alleyway, but to a magical garden seemingly populated with ghosts from the past. But the question is: Who is the ghost: the young woman Tom meets during his midnight wanderings, or Tom himself?

I was especially struck by the scene in this book where Tom realizes that Hatty is a young woman, not the girl he was expecting. I think this happens to parents all the time. We see our children daily, yet one day we look up and they've grown without our even noticing. At our book club meeting yesterday, the moms were looking at a picture taken of the group two years ago, and each of us was startled by how much our daughters had changed.

READING TIME: 3–4 hours, about 229 pages
THEMES: loneliness, passage of time, family, history

Discussion Questions

✦ Why is Tom so reluctant to visit his Aunt Gwen and Uncle Alan? Have you ever visited relatives for several weeks on your own? Compare your experiences with Tom's.

✦ What are Tom's feelings about his aunt and uncle? Do you think he is fair to them?

✦ Dreams often seem to arise from unresolved feelings we experience in our daily lives. In what ways do you think some of Tom's everyday feelings relate to the garden?

- ✦ Why do you think the author chose a garden as the setting for this fantasy rather than something more extraordinary?
- ✦ Tom asks his uncle: "What is Time like, Uncle Alan?" What kind of answer is he looking for? What do you think time is like? Describe experiences you've had where time seemed to go by unusually quickly or slowly. How do some experiences seem to alter the passage of time?
- ✦ Compare Tom's nighttime in the garden to his daytime with his aunt and uncle.
- ✦ There is much discussion of who is the ghost in the book, Hatty or Tom. At any point, did you begin to wonder whether Tom became a ghost every night while Hatty was the live person?
- ✦ What does Tom find so satisfying about the garden?
- ✦ Have you ever had a dream like Tom's, one that was so compelling and pleasurable you couldn't bear to wake up? Share your memories of this dream and explore what features it had that made you long for the dream to continue.
- ✦ Daughters: What do you think this quotation from the book means: "But what can children do against their elders' decisions for them, and especially their parents'?" Do you feel this way? How come? What decisions have been made for you in the past?
- ✦ Mothers: Discuss your interpretations of this quotation from the book: "...nothing stands still, except in our memory." Do you find that your memories "stand still"?
- ✦ Did you anticipate Hatty's identity before the author revealed it?
- ✦ How did you feel when you realized Tom could never return to the garden again? Why? Have you ever had to leave a place forever that was special to you?

ABOUT THE AUTHOR: Philippa Pearce still lives in Great Shelford, England, where she was born and raised. The millhouse where she grew up plays an important part in her books, and her memories and those of her family provide inspiration for many of the scenes. Many say her writing is especially beautiful when read aloud, which Ms. Pearce credits to her 13-year career in radio during which time she developed her feeling for the sounds of words. Ms. Pearce has won many awards for her writing.

Beyond the Book...

CONCEPT DRAWING: The passage of time is one of the major themes of *Tom's Midnight Garden.* Invite book group members to draw a simple representation of how they envision the passage of time. Ask each person to show and explain their drawing. Compare the drawings of mothers and daughters. Does a person's age seem to influence their visual representation of time?

TIME LINE: Make a time line of your life from birth to old age. Illustrate it with photos of times you've already experienced, and drawings or pictures from magazines that represent how you envision your future. Bring your time lines to your dicussion and explain your life portraits to each other.

GARDENING: Do some research on starting a garden. Find someone you can talk to at a local nursery, and ask them to particularly talk about how young gardeners might get started, even if it's just a small widowsill garden. Perhaps they can suggest ways mothers and daughters can garden together. Or start a group garden.

IF YOU LIKED THIS BOOK, TRY...

Tuck Everlasting, by Natalie Babbitt—This is a haunting novel about a young girl's encounter with a family that never grows old (see p. 269).

The Secret Garden, by Frances Hodgson Burnett—The lives of three children are transformed when they discover a weedy, secret garden and bring it to life again.

The Lion, the Witch, and the Wardrobe, by C. S. Lewis—Part of *The Chronicles of Narnia* series, this classic story portrays four children visiting an old house in which a wardrobe leads to a magical land. There are some striking similarities between Lewis's and Pearce's books, particularly regarding the way children experience the passage of time (see p. 166).

Some Other Books by Philippa Pearce:

Fresh: Short Stories

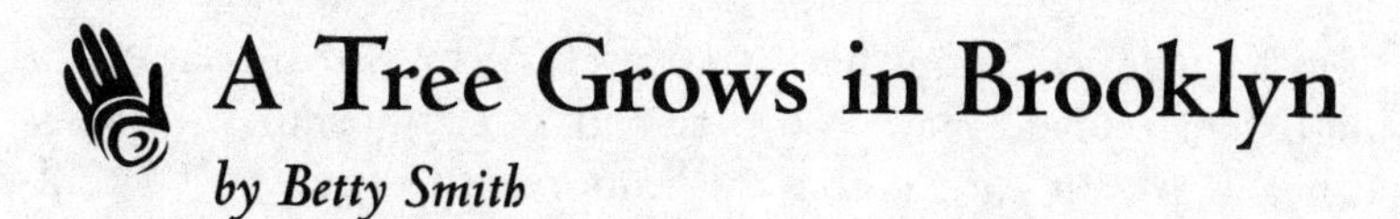

A Tree Grows in Brooklyn

by Betty Smith

A sensitive young girl, growing up in turn-of-the-century Brooklyn, surmounts obstacles of poverty and frustrating family relationships. Ultimately she is able to find her own voice, and make her own way in the world through her creativity, intelligence, and ambition.

Francie's love of books and the important role they play in this novel, really spoke to me. It made me realize how much reading is a part of our life (Morgan still reads in bed with her flashlight, Leroy reads Bible stories and classic books to the girls three or four nights a week, and Morgan and I always read our book club selection together), and how much it means to all of us.

READING TIME: 5–6 hours, about 430 pages
THEMES: conformity, family, self-esteem, creativity

Book Discussion Questions

✦ How does Francie feel about where she lives? Why is her neighborhood important to her?

✦ What kind of relationship does Francie have with her mother, Katie? How does it change over the course of the novel?

✦ How does Francie feel about her brother, Neely? How does she feel about the way her mother treats him? Does that affect the way she feels about her mother? Have you ever felt that way about a sibling?

✦ What's special about Francie's relationship with her father? When Johnny enrolls Francie in a different school so she can get a better education, how does that affect Francie's relationship with Katie? How does the family change?

✦ Francie has a special relationship with books and reading. What makes them so special to her? Have you ever felt this way?

- ✦ How does Francie feel about her Aunt Sissy? Why is this relationship important to her? Have you ever had a relationship with an adult other than your parents that was important to you?
- ✦ Why are the red roses Francie receives at her eighth-grade graduation so important to her? What do they tell you about Aunt Sissy and her father?
- ✦ Why is being an observant Catholic important to Francie? How does it affect her life?
- ✦ Her father's death is a turning point in Francie's life. How does it change her outlook and ambitions?
- ✦ Why does education matter to Francie? How do her feelings about it reflect her family's values?
- ✦ How does Katie's decision that Neely should attend high school and that Francie should continue working affect Francie's relationship with her mother?
- ✦ At the end of the book, the Nolan family's situation is changing because Katie is getting remarried. Is Francie ready to move on? How do we know? How does Francie reconcile her ambitions with the constraints of her situation?
- ✦ What do you think Francie will make of her life?

ABOUT THE AUTHOR: Betty Smith was born Elizabeth Wehner in 1896, and was a second-generation American immigrant. Her grandparents came from Germany to Brooklyn, NY, where she was born and raised. When she married George H. E. Smith, Ms. Smith moved to Ann Arbor, MI, where her husband was a law student. Though she never finished high school, the University of Michigan allowed her to sit in on classes in journalism, drama, writing, and literature. She went on to be recognized for her work in writing drama. In 1934, Ms. Smith left her husband and moved back to New York with her two daughters, Nancy and Mary. Her job with the Federal Theater relocated her to Chapel Hill, NC, where she met and married Joe Jones, a journalist, in 1943. She wrote numerous plays before publishing *A Tree Grows in Brooklyn,* her first novel.

Beyond the Book...

COMING TO AMERICA: For first-generation immigrants, like Francie's parents, the struggle to capture a piece of the American dream was difficult and demanding. Many worked in menial jobs, like cleaning buildings or doing piecework, to support their families. Before the group meets, ask each mother-daughter pair to find out about their own family's experiences in coming to America, and discuss the kinds of work grandparents and great-grandparents had to perform. Share these stories with the group.

BROOKLYN: Do some research into what Brooklyn looked like at the time of the novel, and share pictures and a map of the various neighborhoods and ethnic groups that lived there.

MOVIE: Rent the video *A Tree Grows in Brooklyn,* and talk about the similarities and differences between the novel and the film.

CANDY SHOP: Francie's daily life was punctuated by window shopping or occasional forays to a variety store or small candy shop. In many malls and communities today, there are candy stores that sell old-fashioned treats, or the modern equivalent of the five-and-dime store. Take a trip to one of these stores and picture Francie stopping there on her way home from the library or school.

REFRESHMENTS OR FOOD MENTIONED IN THE BOOK: For Francie and her brother, simple food, like rye bread and sour pickles, were special treats. To evoke the atmosphere of the book, serve simple, sparse refreshments, like rye bread, pickles, and coffee, that convey the deprivation experienced by the characters.

IF YOU LIKED THIS BOOK, TRY...

Mama's Bank Account, by Kathryn Forbes—This is a Norwegian immigrant's memoir about growing up in San Francisco at about the same time as *A Tree Grows in Brooklyn.*

Some Other Books by Betty Smith:

Joy in the Morning

Tomorrow Will Be Better, Maggie—Now

The True Confessions of Charlotte Doyle

by Avi

In this suspenseful thriller, a courageous 13-year-old girl must draw upon all her resources and skills to survive a treacherous sea voyage. In the process, she jettisons many of her most cherished beliefs and assumptions to arrive at an independent identity.

To use Morgan's word, this book was "awesome"! I could not put it down. I'd like to think that Morgan, whose middle name is Charlotta, has all the strength and courage that Charlotte in this book.

READING TIME: 2–3 hours, about 232 pages
THEMES: friendship, loyalty, family, independence, revenge, prejudice

Discussion Questions

- ✦ How does Charlotte feel when she first glimpses *The Seahawk*? Where do these feelings come from?
- ✦ Her intuition tells her one thing, but Charlotte feels compelled to do what the adults tell her to do. Have you ever been in a situation where you felt this way? What was the conflict?
- ✦ Charlotte was raised to be a "proper" young lady, obedient and conventional. How does she experience conflict between the messages she has received from her parents and teachers, and what she finds on *The Seahawk*? Have you ever felt such a conflict?
- ✦ How does Charlotte's self-image as a "lady" initially affect her judgments and perceptions of what happens on board *The Seahawk*? How does that change during the course of the voyage?
- ✦ Charlotte's perceptions, based on her values, cause her to make some serious errors in judgment that have significant consequences. What's an example of how this happens in the book?

Have you ever made such an error? What happened, and what did you learn from these experiences?

- ✦ Why is the dagger that Charlotte receives from the cook, Zachariah, an important gift?
- ✦ How does Captain Jaggery's cruelty towards members of the crew affect Charlotte? What motivates Charlotte to grab the whip, and what are the consequences of her action? What would you have done in that situation?
- ✦ Charlotte has to prove herself to the crew by climbing up the rigging. How does that make her feel? Have you ever had to prove yourself in a similar way?
- ✦ What does justice mean to Charlotte? Do the people around her define it the same way? What does it mean to you?
- ✦ What is the paradox in Charlotte Doyle's liberation from her role and status as a young lady?
- ✦ How does the murder trial affect Charlotte?
- ✦ How does Charlotte's relationship with her family change after her voyage? Have you ever had an experience where you were transformed, and related differently to your family as a result? What was it like, and how did it change you?

ABOUT THE AUTHOR: Avi, whose name was given to him by his twin sister when they were both one year old, was born in Brooklyn, NY, in 1937. Growing up he loved to read, but was better at math and science than he ever was at writing. In fact, he failed so many courses his parents put him in a special school that focused on reading and writing. Yet what eventually made Avi succeed as a writer was not necessarily the extra lessons or attention, but the desire to prove that he could. He has said that the moment you become a writer is the moment "you stop writing for yourself or for teachers and start thinking about readers," and that, "reading is the key to writing. The more you read the better your writing will be." Avi has sons, granddaughters, and a wife, and spends his spare time reading and exploring photography.

Beyond the Book...

CLIPPER SHIPS: Before the group meets, have someone do research into the era of clipper ships, including a map of common shipping and trade routes that would have been comparable to the route that Charlotte Doyle traveled. Someone else could also provide historical background on famous mutinies and shipboard conditions that prevailed during the era, to offer some context for the novel. Some basic information on clipper ships will add to your appreciation of the novel.

MARITIME CENTER: If there is an historical maritime center nearby, tour it and board one of the clipper ships. Explore the Internet to see if tours of Mystic Maritime Center or other places are possible online.

SHIP IN A BOTTLE: As a craft activity, you may want to build a ship in a bottle. These models are easily available at craft or hobby stores.

MOVIE: Rent the video *Mutiny on the Bounty,* a classic movie that depicts a shipboard mutiny in dramatic fashion.

REFRESHMENTS OR FOOD MENTIONED IN THE BOOK: For refreshments during your discussion, serve tea and Scottish shortbread, which Charlotte enjoyed when she was still in the captain's good graces.

IF YOU LIKED THIS BOOK, TRY...

The Slave Dancer, by Paula Fox—This book also describes a grim and dangerous sea journey (see p. 229).

The Witch of Blackbird Pond, by Elizabeth George Speare—A similarly strong-minded heroine has to endure a trial (see p. 302).

Captains Courageous, by Rudyard Kipling—This book shares the same atmosphere and themes as *The True Confessions of Charlotte Doyle.*

Some Other Books by Avi:

Beyond the Western Sea

Nothing but the Truth

Tuck Everlasting

by Natalie Babbitt

Several years ago, our Girl Scout troop saw the play *Tuck Everlasting.* I knew the play was based on a book because my son, Leroy, now 17, had to read it for school in about fourth grade. Well, I found the play so moving, I had to read the book. I was not disappointed. The story is about a bored, restless ten-year-old girl who protects the secret of a shy, gentle family fated to live forever. I found it to be an important book for girls, since the heroine bravely and ingeniously faces painful truths about life, death, separation, and greed. This book is fairly easy to read, but it has some powerful and potentially difficult messages about loyalty and courage.

READING TIME: 2–3 hours, about 150 pages
THEMES: coming of age, mortality, change, greed, loss

Discussion Questions

- What is your impression of Winnie Foster in the first few chapters of the book?
- What does Winnie's treatment of the toad early on, then again later in the book, tell you about the changes she undergoes?
- What are the first hints there is something unusual about the Tucks?
- The Tuck family tells "the strangest story Winnie had ever heard." What is their strange story?
- Compare Miles's and Jesse's feelings about being able to live forever. Which character's feelings are closest to your own?
- The Tucks have some interesting mealtime routines. Is it a good thing to have set routines like mealtimes or not? Why do families have routines at all?
- Compare Winnie Foster's home, the touch-me-not cottage, to the Tucks' old place in the woods. Where would you prefer to live?

- Mae Tuck says she and her family "...don't deserve no blessings—if it is a blessing." Would living forever be a blessing or a curse?
- What would happen if a magical spring promising eternal life were discovered in today's world?
- If you had been entrusted with the bottle of magic spring water, how would you have used it? What do you think of Winnie's decision to pour the magic water on the toad?
- When Tuck brings Winnie out fishing, "Winnie blinked, and all at once her mind was drowned with understanding of what he [Tuck] was saying." What does Winnie suddenly understand about life and death?
- In what ways is life like a wheel or like a stream, as Tuck tells Winnie?
- "At midnight, she would make a difference in the world." How does Winnie's decision to help Mae Tuck escape from jail make a difference in the world?
- Near the end of the book, Winnie's family "sensed that she was different now from what she had been before. As if some part of her had slipped away." How is Winnie different from the girl she once was? What part of her has slipped away?

ABOUT THE AUTHOR: Growing up in Ohio, Natalie Babbitt was surrounded by paints, pens, and pencils, thanks to the artistic encouragement of her mother, who was an amateur landscape and portrait artist. It wasn't until her first author/collaborator, her husband, Samuel Fisher Babbitt, quit one of their collaborations that she first tried her hand at writing. Since then, Ms. Babbitt has written many well-loved books for children and young adults. She is the mother of three and grandmother of three, and now lives in Providence, RI.

Beyond the Book...

ROLE-PLAYING: When Morgan read *Tuck Everlasting* in her fifth-grade class, the kids staged a mock trial and brought the man in the yellow suit back to life. You can do this in your group by assigning some members to play the defense, some to play the prosecution, and others to take on various courtroom roles (Morgan was the Constable). Each girl can spend some time developing her own arguments or lines, and then act them out at the meeting.

CEMETERY: As a mother-daughter outing, visit a cemetery to search for gravestones dated about the same time as Winnie Foster's 1948 gravestone. See if there are names of local families on any of the gravestones. Talk about what it would be like to return to a place to visit someone from your past and discover they were no longer alive.

STREAM: Enjoy a quiet nature experience by sitting beside a stream or river and just thinking about the kinds of life changes Tuck describes in chapter 12. Bring the book along and read favorite passages silently. Perhaps you'll talk about the book's themes on the trip home. If weather permits, hold your discussion outside.

MOVIE: See the 1980 independent film of *Tuck Everlasting.*

REFRESHMENTS OR FOOD MENTIONED IN THE BOOK: The Tucks serve flapjacks, bacon, bread, and applesauce to Winnie for supper, then sit in the parlor rather than around the table to enjoy the meal. For a change, serve these or any other breakfast foods at suppertime, like the Tucks, and enjoy your meal in a different part of the house from where you usually eat.

IF YOU LIKED THIS BOOK, TRY...

The Fall of Freddie the Leaf, by Leo Buscaglia—This comforting book for both adults and children is about the meaning of the cycle of life and death. Its message is very close in spirit to that of *Tuck Everlasting.*

Kneeknock Rise, by Natalie Babbitt—A young boy investigates the causes of some otherworldly goings-on in an evocative setting similar to that of *Tuck Everlasting.*

Some Other Books by Natalie Babbitt:

Goody Hall

The View from Saturday

by E. L. Konigsburg

This winner of the 1997 Newbery Medal is about four sixth-graders who form a record-breaking Academic Bowl team. Each of the members is struggling to grow through a challenge: Nadia's parents are recently divorced, Ethan is forever in his older brother's shadow, Julian has just moved to a small New York town from India and faces hostility, and Noah has a hard time looking beyond his own nose. Their teacher, Mrs. Olinski, has returned to the classroom after a ten-year hiatus recuperating from an automobile accident that left her a paraplegic. Each of the characters brings delightful humor and insight to this quick-paced page turner about the highs and lows of a sixth-grade class and the friendships formed therein.

E. L. Konigsburg is one of my favorite authors because she recognizes and respects the intelligence and creativity of the young adult audience.

READING TIME: 3-4 hours, 163 pages
THEMES: coming of age, race, peer pressure, overcoming handicaps, friendship, inclusiveness

Discussion Questions

- Mrs. Margaret Draper thinks sixth-graders have changed since she started teaching. Instead of saying "Now what?" they say "So what?" Do you think this is true in Mrs. Olinski's sixth-grade class? What kind of change does this description indicate?
- Mrs. Olinski becomes a paraplegic as the result of an automobile accident and has returned to teaching after ten years. How does her physical disability affect her job and what role does it play in the novel? How does she change over the course of the novel?

- ✦ Nadia and Noah tell stories about their summer stays with grandparents in a Florida retirement community. What are your impressions of these senior citizens and their way of life?
- ✦ Why do you think the Saturday tea parties mean so much to Nadia, Noah, Ethan, and Julian?
- ✦ Julian faces racism on many occasions in this novel. What are some of the incidents and how does he respond to them?
- ✦ Julian's father plays a somewhat mystical role as the new owner of Sillington House. What is his background? Do you think it affects his observation of The Souls?
- ✦ Why do you think Nadia chose the name The Souls for their group?
- ✦ Each of the team members tells a story that gives us an idea of who they are and what they bring to The Souls. Nadia, Noah, Ethan, and Julian each face a difficult dilemma in their narrative. Describe these dilemmas. Do they get resolved?
- ✦ What do you think is the meaning of the title?
- ✦ This novel has a very tight structure, though a nonlinear one. What are the different sections and how does suspense build as you read?

ABOUT THE AUTHOR: Elaine Lobl (E. L.) Konigsberg is especially interested in nonconformist children and children with unusual abilities, and is drawn to write about urban settings, especially New York City. She has written and illustrated several picture books and has done ink drawings for many of her novels. Originally a science teacher, she left work to have three children. While at home with them, she became interested in writing and art. She lives on the beach near Jacksonville, FL, and continues to write and draw (see p. 88).

Beyond the Book...

CALLIGRAPHY: Do some research on the art of calligraphy—where it came from, what it's been used for, how it's done. Buy a few calligraphy pens and try your hand at it.

ACADEMIC BOWL: Create your own academic bowl using questions from the book and others you make up.

GRANDPARENT CONNECTIONS: Share stories about your own grandparents—where they live, what they're doing, how they are like or unlike the grandparents in this story.

MAGIC TRICKS: Find some books on magic at the bookstore or library. Have each member of the group learn a different trick to perform at your group meeting. Talk about why magic and the illusion it creates has always been so compelling.

SEA TURTLES: Learn as much as you can about sea turtles and turtles in general. If there's a zoo nearby, go and see the different kinds of turtles there. Plan in advance to speak with the zookeeper, and have her tell you about the animals, and which ones make good pets.

REFRESHMENTS OR FOOD MENTIONED IN THE BOOK: When our mother-daughter book club discussed this book, we did it over "high tea." We got all dressed up—including hats and gloves (largely supplied by one group mom who's a real hat person)—and served tea and biscuits.

IF YOU LIKED THIS BOOK, TRY...

Some Other Books by E. L. Konigsburg:

All Together One at a Time
Jennifer, Hecate, MacBeth, William McKinley and Me, Elizabeth
A Proud Taste for Scarlet and Miniver
Throwing Shadows

Waiting for Anya

by Michael Morpugo

A reclusive widow's farm on the outskirts of a French Pyrenean village is a haven for a growing band of Jewish children hiding from the Nazis during World War II. Although the penalty for aiding Jews is death, the young shepherd, Jo, his father, and the entire village heroically help the children to escape over the border into Spain. Benjamin, their leader, is captured by German soldiers and taken away. The end of the war and the arrival to the village of Anya, Benjamin's young daughter, brighten the book's somber ending.

READING TIME: 2–3 hours, about 172 pages
THEMES: war, good versus evil, race, religion, prejudice, courage, loss

Discussion Questions

- What have you read, heard, or learned about World War II? What do you know about the Holocaust and how Jews were treated in Germany and elsewhere in Europe during the war? What did reading *Waiting for Anya* add to your knowledge or understanding of this period in history? What new ideas or insights did you gain?
- Imagine that you were one of the children in the widow Horcada's barn. Imagine how you might have felt and share those feelings.
- What are your thoughts about the Corporal, and Jo's relationship with him? Do you blame Jo for befriending the "enemy"?
- The Corporal proves himself worthy of Jo's trust and friendship by refusing to search the barn, though he knows that Jews are hiding there. Yet, he is a German soldier and does nothing to save Benjamin and Leah when they are taken away. Do you see the Corporal as a good person? An evil person? Something in between?
- Grandpere accuses himself and the other villagers of being cowards because they did nothing to prevent the Germans from taking Benjamin and Leah away. Do you agree? What should they have done? What could they have done?

- ✦ Benjamin says that when he sees Anya, he will tell her of the kindness and courage of the people who helped him because, "Such things should not be forgotten." People today feel that the Holocaust should not be forgotten either. Why is it important to remember both the terrible and the good things that happened?

ABOUT THE AUTHOR: Michael Morpugo was born in England in 1943. Much of his work is historical fiction. Other recurring themes are how people cope during wartime and stories told from the point of view of animals. In addition to *Waiting for Anya,* a fictional book based on the true experiences of Jewish children during World War II, Mr. Morpugo has written novels about animals, including *War Horse, Little Foxes,* and *Mr. Nobody's Eyes.*

Beyond the Book...

WORLD WAR II RESEARCH: In an encyclopedia, history book, or on the Internet, find out more about World War II. Here are some questions you might think about: What were the causes of World War II? Why did the United States enter the war? In what countries did Jewish refugees settle? Which countries fought on the side of Germany? Which countries fought against Germany? How did World War II finally end? Share what you have learned with your book group.

FRENCH GEOGRAPHY: On a map of Europe, find the part of France near the Spanish border. What is the name of this mountain range there? Do research in the library to find out more about the people who live in the mountain villages in this region.

HOLOCAUST MUSEUM: (see p. 2)

MOVIE: View the film *Schindler's List* to find out more about the Holocaust and how some Germans reacted to it. Afterwards, compare Schindler with the villagers in *Waiting for Anya.*

IF YOU LIKED THIS BOOK, TRY...

Zlata's Diary: A Child's Life in Sarajevo, by Zlata Filipovic—An 11-year-old chronicles life during wartime in 1991.

Four Perfect Pebbles: A Holocaust Story, by Lila Perl and Marion Blumenthal Lazan—Lazan's unforgettable memoir recalls the devastating years that shaped her childhood. Against all odds, she and her family managed to stay together and survive.

The Rescue: The Story of How Gentiles Saved Jews in the Holocaust, by Milton Meltzer—Gentiles rescue their Jewish neighbors from annihilation during World War II.

Some Other Books by Michael Morpugo:

The Ghost of Grania O'Malley

The Wreck of the Zanzibar

Walk Two Moons

by Sharon Creech

Thirteen-year-old Salamanca Tree Hiddle's mother promised to be back before the tulips bloomed, but she's not. When Sal's mother, Sugar, doesn't return as expected, Sal and her eccentric grandparents embark on a six-day journey to find her. Over the course of this multilayered journey, Sal not only unravels the mystery of her mother's disappearance, but comes to understand her own experience in its wake.

Walk Two Moons *is funny and sad, full of mystery, adventure, and even a little romance.*

READING TIME: 3–4 hours, about 280 pages
THEMES: identity, mother-daughter relationships, adoption, loss, death, hope

Discussion Questions

- A mysterious stranger leaves four messages for Sal and Phoebe:

  Don't judge a man until you've walked two moons in his moccasins.
  Everyone has his own agenda.
  In the course of a lifetime, what does it matter?
  You can't keep the birds of sadness from flying over your head, but you can keep them from nesting in your hair.

  Discuss what you think each message means to Sal, the heroine of *Walk Two Moons,* and what they mean to you. Until you learn who actually left these mysterious messages, who do you *think* left them for Phoebe and Sal?

- Why do Phoebe and Sal become friends? How are they alike or different?
- Sal's grandparents believe she is afraid to see her mother. What is Sal afraid of?

- ✦ Why do you think Sal's mother left?
- ✦ How does the author transition between the different stories? What effect does this technique have?
- ✦ How does Sal draw strength from her Native American ancestors?
- ✦ Sal passes the time on the long car journey, en route to her mother, by remembering stories about her past. Describe how your family passes the time on long car trips.
- ✦ Throughout the car trip, Sal keeps hearing words in the wind—*hurry, hurry, hurry,* then *slow down, slow down.* Finally, she again hears the wind telling her to *hurry, hurry* again. What do these "wind messages" tell you about Sal's feelings about the journey towards her mother?
- ✦ Sal's memories of her mother flood back whenever she thinks of blackberries. What foods bring back pleasant memories of your own family or will conjur memories after you've grown up?
- ✦ What are Sal's grandparents like? Describe how Gram decided to marry Gramps. What do you think of Gram's method of judging the suitability of her future husband?
- ✦ Compare Sal's feelings about leaving Kentucky with her Dad's feelings about moving away from their family farm.
- ✦ Ben tells Sal about Pandora's box. Retell the myth in your own words. Describe what the Pandora's box myth comes to mean to Sal.
- ✦ Discuss why Mr. Birkway shows the picture of the vase to his students. What did he want them to learn from this?
- ✦ When Phoebe's mother leaves home, she, like Sal before her, gathers objects that remind her of her mother. What objects remind you of your mother?

ABOUT THE AUTHOR: Sharon Creech won a Newbery Medal for her book *Walk Two Moons* and describes her post-Newbery life as chaotic and exciting. *Walk Two Moons* was inspired by Creech's memories of a childhood road trip as well as the pleasure she finds in nature.

Beyond the Book...

LISTS: Sal gets a mysterious message that says: "In the course of a lifetime, what does it matter?" Consider this question by writing a list of things that matter and don't matter in life. Try to be very specific about your list. For example: Does it matter if you leave dirty dishes in the sink? Does it matter if you forget to kiss each other goodnight? Share and discuss each other's lists.

MAP: On a map, trace the journey Sal, Gramps, and Gram made from Kentucky to Coeur d'Alene and Lewiston, ID. Talk about one special car trip you made—or perhaps one you hope to make in the future.

FORTUNE COOKIE MESSAGES: The little mystery messages Sal receives are very much like fortune cookie messages. Write up a few observations about life, fortune cookie style. Put all the fortunes in a bowl and have people choose one or two. Discuss what each person thinks the writer meant. Then ask the "secret" writer to explain what she actually meant by the fortune.

WALK A MOON: Offer to "wear your mom's shoes" one afternoon at home. Prepare dinner for the family. Help a younger sibling with homework. Tidy up the living room. Think how your mom might "wear your shoes" as well.

REFRESHMENTS OR FOOD MENTIONED IN THE BOOK: The memory of Sal's mother comes alive for her when she remembers the taste of blackberry pie. Suggest that mothers and daughters prepare one favorite family dessert together to share at the book group. Or make a pie together. Enjoy results while talking about family memories associated with certain family foods!

IF YOU LIKED THIS BOOK, TRY...

I Am the Universe, by Barbara Corcoran—A young girl's life changes forever when her mother develops a brain tumor.

The Edge of Next Year, by Mary Stoltz—Two children undergo a painful adjustment to the death of their mother.

Some Other Books by Sharon Creech:

Chasing Redbird

Absolutely Normal Chaos

The Watsons Go to Birmingham—1963

by Christopher Paul Curtis

Ten-year-old Kenny tells about his family, the Watsons of Flint, MI: Momma, Dad, little sister Joetta, and brother Byron, who's 13 and an "official juvenile delinquent." The events that lead up to this family trip to Grandma's in Birmingham, AL, are sometimes hilarious, sometimes moving, and often familiar to just about any child or parent. But the trip culminates in an event that will touch the family's life forever. The 1963 bombing of a Birmingham church, in which four black children were killed. A brief but effective epilogue recounts the main events of the Civil Rights Movement.

I picked this book up when it first came out because my husband is from Flint, MI. We both read it and enjoyed it very much. Leroy said that the characters reminded him of people he knew who were first-generation from the South. Like the Watsons, these people often did not like everything about industrial city living and took or sent their children back home to give them a sense of where their family came from. Both Leroy and I can identify with this; we make sure that our children spend at least two weeks every summer with their grandparents to get an idea about where we came from.

We were also moved by the retelling of the bombing in Birmingham. Last year, at an Equal Rights and Justice exhibition at the Smithsonian Center for African-American History and Culture, artist Radcliff Baily created an installation memorializing the bombing; it included a collage of photographs of the girls who were killed. The exhibit was up for about six months, and the faces of those girls will be ingrained forever in my memory.

READING TIME: 2–3 hours, about 210 pages
THEMES: family, sibling rivalry, friendship, courage, the Civil Rights movement

Discussion Questions

- ✦ Some kids in Kenny's school make fun of him because he's smarter than most and because of his "lazy" eye. Then they pick on Rufus because of his clothes and his Southern accent. How does the teasing affect each of the two boys? How does it feel to be made fun of? How does it feel to make fun of someone else?
- ✦ When Byron beats up Larry Dunn for stealing Kenny's mittens, Kenny says the other kids "wanted to see anybody get it, they'd have been just as happy if it was me or Rufus or someone else." Why do you think the kids in Kenny's neighborhood react this way? Have you ever had a similar experience in which a bunch of kids seem to enjoy seeing someone else "get it"?
- ✦ Byron often seems like a mean kid, and sometimes he acts as though he hates Kenny. How do you know that he has deeper, nicer feelings than those? Why do you think Byron hides his feelings of love and tenderness?
- ✦ Do you think that making Byron stay with Grandma would have been a good way of teaching him how to behave? What are some other ways parents often react to a child who, like Byron, is always getting into trouble? If you were the parent of a child like that, what would you do to try to help your son or daughter?
- ✦ Brothers and sisters have trouble getting along sometimes. Daughters: Do you have one or more brothers or sisters? Do you ever fight or argue? Talk about what causes your disagreements and how you feel when you fight or argue with an older or younger sibling. Mothers: Think back to when you were children and talk about any sibling rivalry you may have experienced. Do you still have any rivalries with your siblings?
- ✦ How does the bombing of the church in Birmingham change members of the Watson family, especially Kenny? Discuss how you felt when you read about the bombing and the children who were hurt or killed. Have you read or heard about similar events that have taken place recently? Why do such things happen? Do you think anything can stop them from happening?

ABOUT THE AUTHOR: Christopher Paul Curtis was born in Flint, MI. He spent his first 13 years after high school on the assembly line of Flint's historic Fisher Body Plant #1. His job entailed hanging doors, and it left him with an aversion to getting into and out of large automobiles, particularly large Buicks. Mr. Curtis's writing, and his dedication to it, has been greatly influenced by his family, particularly his wife, Kaysandra. It was she who launched Mr. Curtis's career by telling him that he "better hurry up and start doing something constructive with his life or else start looking for another place to live." Mr. Curtis has said that for him, "...the highest accolade comes when a young reader tells me 'I really liked your book.'" *The Watsons Go to Birmingham—1963* is Mr. Curtis's first novel, and has won both the Newbery Honor and the Coretta Scott King Honor. Mr. Curtis was found in the public library feverishly at work on his next book the morning the awards were announced.

Beyond the Book...

ANECDOTES: One of things that makes *The Watsons Go to Birmingham—1963* fun to read is the amusing anecdotes, or little stories, that Kenny tells about his family; for example, the time Byron's lips froze to the sideview car mirror, or the story about the "Ultra-Glide." Can you remember a funny incident involving members of your family? Try writing about such an incident. Share your anecdotes with each other.

CIVIL RIGHTS MOVEMENT: At the end of the book, the author gives you a brief overview of the Civil Rights Movement and mentions some of its heroes. Choose an important event during the Civil Rights Movement, or an important person such as Dr. Martin Luther King Jr., or Rosa Parks, and find out more about this event or person. Use an encyclopedia, books in the library, or the Internet. You may want to read a biography. Share your findings during your book discussion.

MOVIE: Spike Lee made a documentary about the bombing in Birmingham called *Four Little Girls.* Watch it and compare his portrayal of the bombing with Christopher Paul Curtis's.

REFRESHMENTS OR FOOD MENTIONED IN THE BOOK: My husband loves to drive, and every time we take a long trip I am responsible for the food. We end up eating lots of sandwiches like the ones they have in this book. Serve a platter with bologna, tuna fish, and peanut butter and jelly sandwiches. Accompany it with some potato salad, soda pop, and fruit.

IF YOU LIKED THIS BOOK, TRY...

Jacob Have I Loved, by Katherine Paterson—This Newbery Medal-winner, set in the Chesapeake Bay region, focuses on the rivalry between two sisters (see p. 153).

Freedom's Children: Young Civil Rights Activists Tell Their Own Stories, by Ellen Levine—Southern African-Americans, who were young and involved in the Civil Rights Movement during the 1950s and 1960s, describe, firsthand, the Montgomery bus boycott, the integration of schools, the Selma-to-Montgomery march, and more.

Some Other Books by Christopher Paul Curtis:

He Looks This Way

The Westing Game

by Ellen Raskin

Sixteen residents of the Sunset Towers apartment complex find out mysteriously that they are potential heirs to a million-dollar fortune. To receive the money, though, they must work in pairs to find out which of the 16 residents murdered their benefactor, Sam Westing. But as the book goes on, the mystery of the billionaire's death becomes far less important to the potential heirs than what they learn about each other, and ultimately about themselves.

When our group covered this book, we did something we had never done before: we invited the dads. We chose this book because it had an element of problem-solving, something we though most dads would enjoy. We were right. All of the dads came to the meeting (even one who had to travel from New Jersey), and all but one of them had read the book. During the meeting, we noticed that dads tend to lecture more than discuss, and that each is sure to espouse their knowledge of the events of the book, and let it be known how much they had studied the characters. When this went on too long, or when the discussion became dominated by the parents, the girls would jump in and reclaim it. It was interesting to note that some girls who usually tend to be quiet spoke up more with their dads in the room, while others had less to say than usual. Afterwards, almost all of the dads invited themselves back, saying that they had a wonderful time, and though they hadn't known what to expect, they really enjoyed themselves. But while it was fun once, we don't intend to invite them back.

READING TIME: 3–4 hours, about 217 pages
THEMES: revenge, greed, disability, gender roles, prejudice

Discussion Questions

- Without giving away the ending, share with other readers your summary of the plot.

- ✦ Describe the strange setting of the book. What particular details let you know that something mysterious is about to happen?
- ✦ There are 16 characters who share clues to the mystery of Sam Westing's disappearance. With which character did you most identify? Why?
- ✦ One at a time, describe how you put together the clues as you read along. Did other readers put the clues together differently than you did?

This book's mystery within a mystery is tremendously engaging, and I plan to read it again and again.

- ✦ What was interesting about the way Sam Westing paired up people to investigate the clues? Describe how each of the eight couples went about figuring out their clues. Compare the young people who are heirs to Sam Westing's fortune to the older heirs. In what ways do the younger characters approach the mystery differently than the older characters?
- ✦ What is Turtle Wexler like? Would you have been bold enough to enter the Westing house on Halloween night as Turtle did? Have you ever gone into a scary place just because you were curious?
- ✦ Some questions about the role of women revolve around the bride-to-be, Angela. Compare Turtle and Angela. What do you think the author is saying about the position of women in society?
- ✦ What kind of person was Sam Westing? What clues led you to your conclusion about him?
- ✦ How soon did you figure out how the book would end? Were you surprised or disappointed?

ABOUT THE AUTHOR: Ellen Raskin has written and illustrated many books. She uses her visual and writing talents to create stories which she hopes will be full of surprises for her readers. She believes that books can be fun and tries to pass this message along. Ms. Raskin lives in New York City with her husband, Dennis Flanagan, and her daughter from a previous marriage. She has won numerous awards, both for her writing and her illustrations.

Beyond the Book...

AMERICA, THE BEAUTIFUL: Before your book discussion, obtain the complete lyrics to the patriotic song, "America, the Beautiful." Using the lyrics, figure out which clues are related to the song. Consider why Mr. Westing chose that song as part of the mystery.

CLUE: Play a game of "Clue" with mother-daughter pairs or mothers versus daughters. Players should note how they hide or reveal their cards by their facial expressions, comments, and other behavior.

PARTY: Host a murder mystery party (kits are available in party stores) in conjunction with your book discussion. Perhaps you can "set up" the "crime" as a surprise for family, friends, or book club members to solve. Invite players to work together as pairs while figuring out the mystery. Afterwards, discuss how each helped the other in trying to solve the mystery.

REFRESHMENTS OR FOOD MENTIONED IN THE BOOK: To capture the atmosphere of the book, create a Halloween setting for your book discussion of *The Westing Game.* Serve Chinese food and Halloween candy to add to the occasion.

IF YOU LIKED THIS BOOK, TRY...

The Bodies in the Bessledorf Hotel, by Phyllis Reynolds Naylor—Bodies start disappearing in the hotel where a boy's father works. This strange hotel has a similarly peculiar atmosphere to Sunset Towers.

The House on Hackman's Hill, by Joan Lowery Nixon—Two children are visiting a haunted house when they are trapped by a snowstorm.

Goody Hall, by Natalie Babbitt—Some humorous and mysterious events are going on in the Goody family mansion.

Some Other Books by Ellen Raskin:

Figs and Phantoms

Spectacles

Nothing Ever Happens on My Block

Where the Lilies Bloom

by Vera and Bill Cleaver

Fourteen-year-old Mary Call Luther becomes the head of her family when she, her brother, and two sisters are orphaned in the mountains of Appalachia. Bound to a promise she made to her father to keep the family together no matter what, Mary Call struggles to eke a living out of the mountains by "wildcrafting," gathering medicinal plants. To stay together, the children conceal the death of their father and struggle to hang on to their tumbledown farm.

READING TIME: 2–3 hours, about 210 pages
THEMES: death, poverty, adversity, independence, survival, family

Discussion Questions

- What are the promises that Roy Luther forced Mary Call to make?
- Discuss whether you think it's fair for a parent—or anyone—to pressure another person, particularly a young person, into making binding promises.
- When the moment of Roy Luther's death arrives, Mary Call isn't sure she can handle the situation. She says this prayer: "Oh, Lord, You sure made a mistake when You put me together. You didn't give me enough strength to carry out all I'm supposed to do." Discuss whether Mary Call's prayer is a true assessment of her strength or not.
- What were Mary Call's reasons for concealing Roy Luther's death from Kiser Pease and the Connells? Do you think she was right or wrong in keeping his death a secret for so long?
- Do you agree with what Roy Luther told Mary Call: "...that there would never be any understanding of death, that it was beyond people's notions and ideas and meant to be that way"?
- Discuss your image of Roy Luther from what his children say about him at his burial site.

- ✦ Fourteen-year-old Mary Call deals with her family's countless setbacks with cleverness and foresight. Give examples of these traits.
- ✦ Mary Call is often hard on herself. At one point she says: "...because I am a pessimist and must always keep sticking my tongue in pessimism the way you do a sore tooth I couldn't help thinking that it was all too easy. Things just aren't easy for people, I said to myself." Do you think of Mary Call as a pessimistic person?
- ✦ Discuss Mary Call's observation that the purpose of everyone's life is "...what they make of being here."
- ✦ Discuss the incredible "onion cure" the Luther children administer to Kiser Pease.
- ✦ Mary Call refuses any kind of charity available to her family. She says: "Charity is one of the worst things there is. It does terrible things to people. ...It demeans people." What does Mary Call mean by this? Do you agree or disagree with her view that charity is "one of the worst things there is"?
- ✦ *Where the Lilies Bloom* takes place over several seasons. When spring returns to the mountains, Mary Call greets it with this comment: "Spring is a wondrous necessity." What does Mary Call mean by this? What is your favorite season?

ABOUT THE AUTHOR: Vera and Bill Cleaver wrote many short stories and 16 novels for children and young adults. Together they would mull over an idea and establish the details, and then Vera would write the book. This process continued until Bill's death, after which Vera continued to write alone. Though they came from very different backgrounds, Bill and Vera made a great writing team. Bill's military career with the U.S. Air Force enabled the couple to travel widely, an opportunity that influenced their writing significantly.

Beyond the Book...

HOME REMEDIES: The Luther children earn small amounts of money by "wildcrafting" plants and roots for medicine. Since the book was written in 1969, home remedies made from wild plants have become more popular and accepted by the general public. Obtain a booklet from a health food store about some of these home remedies and what they claims are made for. Speak to someone who has tried some of these home remedies or discuss any special home remedies passed on to your family from an earlier generation.

EDIBLE FLOWERS: With the help of a naturalist or field guide, research and locate some edible flowers like red clover, wild rose, or wild strawberries. (Eating or using wild plants can be dangerous. If you are at all unsure about the species you are identifying consult other texts or a botanist.)

MAP: Appalachia remains one of the poorest regions of the country—and its people one of the most independent and private. Point out the area on a map. Brainstorm ways to help people in this area—perhaps by donating money, clothing, or food, or by working in some way with Habitat for Humanity, which helps repair and build houses for families in Appalachia. Imagine running a household without regular electricity, no hot water, a caved-in roof, hand-me-down clothes, and shoes that don't fit.

MOVIE: See the 1974 movie *Where the Lilies Bloom.*

REFRESHMENTS OR FOOD MENTIONED IN THE BOOK: The Luther children know how to make a good meal from a ham hock, a little bit of flour and butter for biscuits, and wild greens. Find some recipes for these dishes and serve them at your book discussion.

IF YOU LIKED THIS BOOK, TRY…

M. C. Higgins the Great, by Virginia Hamilton—The story tells of a black Appalachian family whose home is threatened by a mountain slide. A Newbery Medal winner.

Where the Red Fern Grows, by Wilson Rawls—This story, set in the Ozark Mountains, tells of a determined young boy coming of age in a rural community. It is similar in feeling and spirit to *Where the Lilies Bloom* (see p. 292).

Some Other Books by Vera and Bill Cleaver:

Delpha Green and Company
I Would Rather Be a Turnip
The Kissimmee Kid (by Vera Cleaver)
The Mimosa Tree

Where the Red Fern Grows

by Wilson Rawls

In this heartbreaking Depression-era novel, a ten-year-old Ozark Mountain boy, Billy, spends two years earning enough money to buy two hunting dogs for fur trapping. When the puppies finally arrive, the boy devotes all his time to their training, and they soon become the best coon dogs in the area. The bond between the boy and his animals is so profound that when the dogs die after a valiant fight with a mountain lion, the grief nearly overwhelms him. Ultimately their deaths are the making of his manhood.

Billy's story reminded me of my own son, Leroy, and how, when he was nine, he worked to earn money to buy his own dog. This book is a reminder of the satisfaction we get from working for something we really want.

READING TIME: 3–4 hours, about 250 pages
THEMES: relationships with animals, hard work, family, grief, coming of age

Discussion Questions

- Describe Billy's deep longing for two hunting dogs. Why does the idea mean so much to him? Have you ever had this tremendous longing for a pet?
- Billy works for two years to save enough money for his hunting dogs. How and why does that change the experience of getting them? If you have a pet, share your memories of choosing the pet and coming home with it.
- Billy's dad defines the feelings between humans and dogs in this way: "...I call it love—the deepest kind of love." Discuss whether you think the bond between a person and a dog is love or loyalty.
- The bond between Bill and Old Dan and Little Ann is intense. Why are his relationships with them so strong? If you have a pet,

describe your relationship with your own pet. How is it different from Billy's with his dogs?

- How does the Depression affect how Billy's family lives? What would a depression change about today's society?
- Billy and his sisters—as well as his mother and father—have very specific, gender-based responsibilities in their rural, Depression-era home. At one point, Billy says: "After all, she was a girl, and girls don't think like boys do." Discuss this quotation in relation to the book and to your own life. Do you agree or disagree with this point of view?
- At times Billy's grandfather seems to understand his grandson better than Billy's own parents do. Describe any relationships you have with adult relatives or family friends in which you and that person maintain a special understanding of each other, separate from your parents.
- Hard work and determination are two major subjects in this book. Billy's grandfather scolds his grandson at one point, saying: "Don't ever start anything you can't finish." Do you think that's good advice? Some people might argue that abandoning a doomed plan can be a good idea too. Discuss the pros and cons of sticking with ideas or projects no matter what.
- Raccoon hunting was a way for poor country people to make money during the Depression. Did the subject of hunting raccoons bother you as you read about it? If so, did the author handle the subject in a way that helped you to accept it? How so?
- Why doesn't Billy want his dogs to kill the "ghost coon"?
- How did you react to the deaths of Old Dan and Little Ann?
- Talk about some of the scenes that had the most powerful effect on you. Why did they make such an impact?
- Billy's father sees the deaths of Old Dan and Little Ann as a sign from God that Billy is to join his family in town and not stay behind to continue hunting. Do you think that everything that happens is part of a master plan?
- Why does Billy feel so strongly that he must stay with his dying dogs and bury them alone?

✦ What is the significance of the red fern? Why do you think the author chose this title?

ABOUT THE AUTHOR: Before he encountered actual books, author Wilson Rawls was creating stories that he often told to his constant companion, his dog. Growing up in the Oklahoma Ozarks, Mr. Rawls spent his youth in the heart of the Cherokee nation. When his family moved to Muskogee, he was able to attend high school. *Where the Red Fern Grows* has become a young-adult classic and has been made into a motion picture.

Beyond the Book. . .

RACCOONS: Billy and his family deeply respect the intelligence of raccoons and do not wantonly hunt them. Research information about raccoons to share with each other as part of your discussion. Also, talk about hunting in general—whether or not the idea of it is objectionable to you, what makes some hunting more acceptable, and so on. Investigate whether hunting is practiced in your area. If so, find out what the hunters hunt and what they do with the animals they kill.

DEPRESSION: Research rural life during the Depression to understand the context for Billy's story.

SAVING: Mothers and daughters may want to brainstorm ways the girls can save for something worthwhile that seems out of reach. Remember that Billy's dogs had all the more value because he worked and waited so long for them. Or prepare a budget and save for something together.

NATURE CENTER: Go to the local nature center to see animals that have been injured in the wild.

PET VISIT: Organize a pet-visiting day for people who are in institutional settings like hospitals or nursing homes.

MOVIE: See the 1974 movie version of *Where the Red Fern Grows.*

REFRESHMENTS OR FOOD MENTIONED IN THE BOOK: In Depression-era times, food was pretty basic in poor homes like Billy's. On one benchmark day, Billy's grandfather gives him a bag of peppermint sticks, jawbreakers, and gumdrops. You may enjoy sharing these old-fashioned "penny candies" during the discussion of this book.

IF YOU LIKED THIS BOOK, TRY…

Old Yeller, by Fred Gipson—A stray dog wanders into the life of a Texas farm boy, and a deep bond develops between the two.

Rascal, a Memoir of a Better Era, by Sterling North—This book tells of the adventures of a young boy and his pet raccoon.

Some Other Books by Wilson Rawls:

Summer of the Monkeys

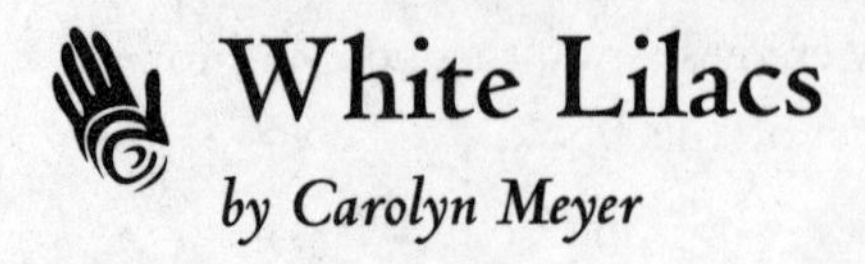

White Lilacs

by Carolyn Meyer

A 12-year-old, African-American girl grows up quickly when her white neighbors decide that "people like her" no longer belong, and attempt to move the black people in the community out of their homes. How she fights back on her terms and retains her dignity and integrity is the heart of this engrossing story.

READING TIME: 2–3 hours, about 237 pages
THEMES: racism, self-esteem, identity, friendship, family, values, loyalty

Discussion Questions

- ✦ How does Rose Lee feel about working for the Bells?
- ✦ How does becoming part of the Bells' household staff that summer affect Rose Lee's friendship with Catherine Jane?
- ✦ How does Rose Lee feel when the white people talk about her neighbors and other black people as if she weren't there?
- ✦ Her brother, Henry, resents the white community, and doesn't want to work for whites in their homes as the slaves did. How does his attitude influence Rose Lee? What position do her parents and grandparents take on the issue? At this time, and in this world, what does it mean to "know your place"? Do you ever feel like you're expected to "know your place"? How does, or would, that feel?
- ✦ Why is Miss Emily Firth an important person in Rose Lee's life?
- ✦ When her Aunt Susannah comes to live in Freedomtown, she opens up a world of new possibilities, ambition, and poetry for Rose Lee. How does Rose Lee's life change as a result? Have there been any important adults in your life, besides your parents, who have exerted a comparable influence on you?
- ✦ Many of Rose Lee's neighbors were reluctant to believe that they

would be forcibly moved from their homes. What happens after the Ku Klux Klan marches into the black community and burns a cross in front of their church?

- How does Rose Lee feel when she finds out that Tom Bell is a member of the Ku Klux Klan?
- As Rose Lee is serving dinner one night, she hears the town's mayor talk about getting rid of her friends and neighbors. She thinks to herself, "showing pride in Dillon means getting rid of my family and friends, all of us." What does this mean to Rose Lee?
- What's important about Miss Firth's parting gift to Rose Lee?
- Flowers, plants, and growing things are especially meaningful in Rose Lee's family. What's significant about the white lilacs that her grandfather plants and gives to her? What do they represent?

ABOUT THE AUTHOR: Carolyn Meyer began writing when she discovered her father's diary shortly after his death. Learning of his aborted love affair inspired her to create her first book. She usually forms her characters first, from her experiences and the lives of the people she knows, then she creates scenarios around them. In addition to *White Lilacs,* she has written almost 40 fiction and nonfiction books.

Morgan read White Lilacs *in her fifth-grade class and liked it so much, she suggested that I read it and include it in this book.*

Beyond the Book...

PRESENTATION: Displacement of unpopular ethnic groups from communities didn't just occur to black Americans during the post-Civil War and Jim Crow eras. Before your group meets, have some members research other situations that were similar to what these fictional characters experienced, whether it was the situation faced by the Irish in the 19th century, the Jewish peasants who were uprooted by the Russian Czar at the turn of the century, the Japanese-Americans' internment during World War II, or prejudice against Cambodian or Laotian immigrants after the Vietnam War. Open a discussion on how it would feel to be unwanted by neighbors because of your religion or skin color.

JIM CROW: Although slavery was ended and the Civil War was a memory, the impact of the Jim Crow laws (segregation on train cars, in schools, at drinking fountains, and so on) pervaded daily life for African-Americans. Research Jim Crow laws and what they meant to the people living under them. Someone may want to prepare a map of Southern blacks migration to Chicago and other Northern cities (or other places where life was easier) as a result of these laws.

JUNETEENTH CELEBRATION: The Smithsonian's Anacostia Museum of African-American History and Culture has an annual Juneteenth celebration, which serves to remind the public of the day, June 19, 1865, Union Army soldiers arrived in Texas and told the slaves that they were free. Today many African-Americans come together with celebration traditions from the early days like parades, picincs, music, speeches, and dances. Create your own Juneteenth celebration and invite family and friends; it can be as big or as small as you like.

LIFT EVERY VOICE: Have someone bring in a copy of the lyrics to what has been known as the African-American National Anthem, the hymn "Lift Every Voice and Sing," and discuss why that song has taken on such an important role.

REFRESHMENTS OR FOOD MENTIONED IN THE BOOK: To convey the spirit of the Juneteenth celebration, a pivotal event in the book, serve refreshments like ham, biscuits, corn bread, fried chicken, sweet potato pie, and pecan pie.

IF YOU LIKED THIS BOOK, TRY...

Any of Mildred D. Taylor's books would be a logical follow-up to this novel (see p. 104).

I Know Why the Caged Bird Sings, by Maya Angelou (see p. 142).

Some Other Books by Carolyn Meyer:

Where the Broken Heart Still Beats

In a Different Light

Wise Child

by Monica Furlong

In this fantasy novel, a young girl named Wise Child is orphaned and sent to live with a single woman named Juniper. This new caretaker, knowledgeable in the ways of common-sense healing as well as in matters of the spirit, is deemed a witch by superstitious countryfolk and religious elders. But Juniper teaches Wise Child common-sense living, the practical knowledge of healing plants, intellectual rigors of Latin and mathematics, and the understanding of how all human, plant, and animal life is connected in what Juniper calls "the pattern." Most of all, Juniper teaches Wise Child the power of love over ignorance and greed.

I loved the way this book makes the reader focus on the intrinsic value of ordinary tasks, such as cooking, cleaning, gardening, and studying. My mother used to tell me that cleaning is good for the soul, and we spent some of our best times together cleaning the house. *Wise Child* makes you reach inside yourself and ponder the things that are truly important in life.

READING TIME: 3 hours, about 228 pages
THEMES: identity, learning, healing, sorcery, temptation, hard work, coming of age

Discussion Questions

- ✦ Why does Juniper's community fear her? How is she a threat to them?
- ✦ What do you think of the "auction" of Wise Child to the villagers? Why does Wise Child wind up living with Juniper?
- ✦ Compare how Juniper is perceived by the countryfolk with how she appears to you while raising Wild Child.
- ✦ Compare Juniper to Wise Child's real mother. In what ways does

Juniper embody the qualities of an ideal mother? Would you like to have had Juniper as a mother?

- ✦ What do you think of Wise Child's father? Do you think he should have taken Wise Child with him and raised her on a ship?
- ✦ Do you think Juniper is a witch? If so, why? Why do you think witches have been feared throughout history? Do you believe in witches? Consider whether they possess supernatural powers or possibly another way of looking at the world.
- ✦ Juniper has an interesting view of cleanliness, order, and housework. She says to Wise Child: "...as you clean house up, it gives you time to tidy yourself up inside—you'll see." What does she mean? Would you agree that cleaning your outer space somehow enables you to put your inner space in order as well?
- ✦ How does the country community discourage Wise Child and other girls from being independent? In what specific ways does Juniper encourage Wise Child's independence? Discuss whether you think today's girls and women, like Wise Child and Mauve, are tempted away from work, knowledge, and the development of skills by cultural promises of ease and luxury. Consider how such temptations are communicated to today's females.
- ✦ Wise Child arrives at Juniper's cottage with dozens of superstitions. Talk about some of these, then discuss how Juniper explains them away with common sense. Why do you think superstitions develop at all?
- ✦ Discuss what you think of the idea of a *doran:* "It was someone who had found a way in to seeing or perceiving...the energy—the pattern." Juniper explains *dorans* can be poets, singers, craftspeople, healers, or people who "just know." Do you know anyone like this?
- ✦ Discuss how Wise Child comes to embody her name by the end of the book.

ABOUT THE AUTHOR: Monica Furlong is an acclaimed author of young-adult books. She is also known for her biographies of prominent spiritual figures, including Alan Watts, Saint Thérèse of Lisieux, and Thomas Merton. Ms. Furlong currently lives in London, England.

Beyond the Book...

SUPERSTITIONS: Make a list of superstitions you've heard of or believed in. Take a book on the topic out of the library to learn where these superstitions may have originated and what knowledge, experience, or common sense might explain them away.

HERBAL MEDICINE: Investigate medicinal plants mentioned in the book and perhaps grow some of them together to use in making remedies for minor physical ailments. Mothers and daughters might want to talk about ways to avoid too much dependence on prescription medications and investigate nonmedical ways of healing simple ailments.

CLEANING: Set aside an afternoon to do some cleaning together. Put on some music, have good snacks around, and scrub away. When you're done with the area you've chosen, sit down and look around. Talk about how it felt to be cleaning, what you thought about, and how you feel now that it's done.

REFRESHMENTS OR FOOD MENTIONED IN THE BOOK: Serve peppered cheese or honey on dark bread. Make some herbal tea as well.

IF YOU LIKED THIS BOOK, TRY...

Juniper, by Monica Furlong—This prequel to *Wise Child* tells the story of young Juniper, who comes to believe that she has special powers.

The Midwife's Apprentice, by Karen Cushman—This Newbery Medal–winner, set in medieval times, is also about an orphaned girl empowered by an older woman who is wise in the ways of the healing arts.

Some Other Books by Monica Furlong:

Robin's Country

The Witch of Blackbird Pond

by Elizabeth George Speare

Kit Tyler, an orphaned teenager from Barbados, is sent to live with strict relatives among the Puritans of New England. In this engrossing historical novel, she learns to adjust to her newfound family, a challenge that ultimately strengthens her character.

READING TIME: 3–4 hours, about 250 pages
THEMES: friendship, loyalty, family, love, choices, values, conformity, identity

Discussion Questions

- ✦ How does Kit feel about leaving Barbados? Why? How would you feel about leaving your home?
- ✦ When Kit swims to rescue Prudence's doll, she is called a witch. Why? How does that make her feel about the community she's about to join?
- ✦ New England isn't exactly what Kit expected, nor is the Wood family's home. How does she react to her new surroundings? Have you ever been in a similar situation? What was your reaction?
- ✦ Kit arrives at the Woods' bearing elegant and expensive gifts for Judith and Mercy. How does Kit feel when Aunt Rachel won't let them accept the gifts? How does that rejection affect her relationship with Uncle Matthew? What about her relationship with Aunt Rachel?
- ✦ What kinds of things are important in the Wood household? Do they share these values with their Puritan community? In the household, who best reflects these values—Aunt Rachel, Judith, or Mercy?
- ✦ What kind of relationship does Kit have with her cousin Mercy? What about with her cousin Judith? How are Judith and Mercy similar? How are they different? Which of the cousins do you like better, and why?

- ✦ Why does Kit continue to see Hannah, the shunned Quaker woman, despite her uncle's disapproval?
- ✦ Why is Prudence important to Kit? Why is Hannah's friendship important to Kit? How is it different from her relationship with Prudence?
- ✦ Why does Kit accept the attentions of William Ashby?
- ✦ How does Kit feel about Nat at the beginning of the book? How does their relationship change?
- ✦ How does the trial affect Kit? How does she change as a result of this experience?
- ✦ What do you think about the couples that form at the end of the book? Are the three pairs well-matched? Why or why not?

ABOUT THE AUTHOR: Elizabeth George Speare was born in 1908 in Melrose, MA. She was educated at Smith College and Boston University, and then went on to become a teacher. She published her first novel in 1957, and has since written a number of well-loved and highly acclaimed books for children and young adults. She has been recognized as a leading author of historical fiction, and has won both the Scott O'Dell Award for Historical Fiction and the Laura Ingalls Wilder Award for her enduring contributions to children's literature. Ms. Speare died in 1994.

Beyond the Book...

HANDMADE: Try making your own soap, or doing some embroidery. You can usually find kits for either at a hobby or craft shop, or find the recipe and instructions in a craft book. While you're making your soap or embroidery, talk about how it would affect your life if you had to make everything you need by hand. Do you feel differently about washing your hands with soap you made than with soap you bought in the store? Why?

RESEARCH: Gather some books on Puritanism, Quakers, and witchcraft. Imagine living in a town with these values. Compare our current lifestyle with the way people lived in Puritan communities. What are the nice things about having those restrictions or rules in your life? What are the negative aspects?

MOVIE: Either of the two film versions of *The Crucible*—about the Salem witch trials—will give you a sense of life in a Puritan community during this time period.

REFRESHMENTS OR FOOD MENTIONED IN THE BOOK: In their everyday meals, Kit's Puritan family eats a lot of things made with corn. For your discussion, prepare one of the foods they mention—corn pudding, corn bread, corncake—and talk about why corn was such a staple in their diet. Or re-create the wedding feast by making apple, mince, and dried-berry pies; little, spiced cakes with maple-sugar frosting; candied fruits and nuts; and sweet apple cider.

IF YOU LIKED THIS BOOK, TRY...

The Crucible, by Arthur Miller—This story deals with the Salem witch trials. You might also try Miller's *Break with Honor,* an historical novel about the same era.

Some Other Books by Elizabeth George Speare:

The Bronze Bow

The Sign of the Beaver

The Wolves of Willoughby Chase

by Joan Aiken

When Bonnie Green's parents leave on a sea voyage, Bonnie and her cousin Sylvia find themselves in the care of the evil governess, Miss Slighcarp. While under her supervision, they discover that Miss Slighcarp and her wicked accomplice, Mr. Grimshaw, are plotting to take over the Green family's fortune. With the help of Simon the Gooseboy, Bonnie and Sylvia escape the cruel clutches of Slighcarp and Grimshaw, and foil their sinister plans.

Morgan read this book in her fifth-grade class and had been raving about it and telling me to read it ever since. I finally did and enjoyed it very much. While Morgan likes the mystery and intrigue of the story, I am more impressed with how the author handles the theme of good versus evil. This book is sure to start an important discussion about human nature.

READING TIME: 2–3 hours, about 168 pages
THEMES: good versus evil, courage, assertiveness, generosity, loyalty

Discussion Questions

- Would you call *The Wolves of Willoughby Chase* a realistic story, or is it more like a fairy tale? Why?
- In most fairy tales, characters are exaggerated so that they are either purely good or purely evil, with the good ones winning out in the end. Are characters so one-sided in most realistic stories, or do most of them have a combination of good and bad traits? Do realistic stories always have happy endings? Why do you think this is so?
- Who are the good characters in this book? The evil characters? Talk about each character and what makes her or him good or evil.
- In the story, which traits of Bonnie's does Miss Slighcarp criticize

the most? Which traits help the children to survive and win out in the end? Have you ever had any experiences where being assertive rather than quiet and obedient helped you succeed? Discuss such experiences with your group.

- Real life is not as simple as life in fairy tales. Most people are not purely good or bad, but some combination of the two; and not every ending is happy. Think about people you know as examples—including yourself. Do you think most people are mostly good or mostly bad? Do you think good wins out more than evil?

- An author of a story or book often has a main purpose in mind when writing. Some main purposes are: to teach, to entertain, to persuade readers to think or act in a certain way, or to stir the readers' emotions. What do you think Joan Aiken's purpose is in *The Wolves of Willoughby Chase?*

- Did you find any part of the story suspenseful or scary? Do you sometimes like to be kept in suspense or scared by books or movies? If so, why do you think you do? How do you feel at the end of a story that kept you in suspense or scared you? Why is it often fun to be scared by a story, but seldom fun to be scared in real life?

- Think about the way Joan Aiken names her evil characters. How do even their names make them seem evil? What about the name of the town where the school for orphans is located?

- Think about the title of the book. Who are the real "wolves" of Willoughby Chase?

ABOUT THE AUTHOR: Joan Aiken has written over 80 books for children and adults. She was born in Rye, East Sussex, in England, and was home-schooled by her mother until she was 12. Ms. Aiken says she began composing stories on solitary walks in the country, and when her younger brother would tag along, her stories became more elaborate in order to entertain him. Ms. Aiken's books never sugarcoat the difficulties of life. Her first husband died of lung cancer and she raised her two children alone. Ms. Aiken's own children are her best critics and she reads to them often throughout the writing process.

Beyond the Book...

STORYTELLING: Pick out a few fairy tales you remember from when you were small and tell one or two of them to younger siblings or other small children. Try to choose stories that have scary parts, such as *Little Red Riding Hood* or *Hansel and Gretel.* As you tell or read one of the stories, be aware of your audience's reactions. Compare the childrens' reactions to scary fairy tales with your own reactions to *The Wolves of Willoughby Chase.*

WRITING: Make up an original story in which a conflict between characters is really a conflict between good and evil. Give your story a happy ending, and give your characters names appropriate to their goodness or "evilness." Read your story to your family or book group. Invite members to act it out.

CHARADES: When Morgan read this book in school, her class took some of the more interesting vocabulary words, looked them up, and played a game of charades with the new words they had learned.

REFRESHMENTS OR FOOD MENTIONED IN THE BOOK: There's plenty of delicious food to re-create from this book. For a light lunch, make a potato soup like the one the kids in this book were served, or oyster patties. For a fuller meal, make roast partridges with bread sauce, red currant jelly, and vegetables, or have some sliced roast beef. For dessert, serve lemon tarts, apple cheesecake, or plum pudding.

IF YOU LIKED THIS BOOK, TRY...

Is Underground, by Joan Aiken—This is another episode in *The Wolves Chronicles.* Other scary stories by this author are: *Give Yourself a Fright, A Fit of Shivers,* and *A Creepy Company.*

The Black Cauldron; The Book of Three; and *The High King,* by Lloyd Alexander—These three titles form a trilogy of humorous fantasy in which the powers of good win out over evil.

Some Other Books by Joan Aiken:

Cold Shoulder Road
The Angel Inn
Midnight Is a Place

The Woman in the Wall

by Patrice Kindl

This almost-fantasy novel is about the withdrawal of an extremely shy girl from family life. Anna's father has abandoned her, her two sisters, and her mother. Anna's mother, now a single parent, is busy keeping the household together and doesn't notice Anna's increasingly bizarre shyness. Anna becomes so reclusive, she creates a parallel, solitary life for herself in secret chambers she has built within the family's rambling house. Her emergence, years later, back into family life, is a triumph.

Thank God for our mother-daughter book club! That's the thought that kept going through my head while reading this wonderful book. At some point during adolescence all girls try to hide in some way, but a mother-daughter book club may actually keep that from happening. With a club such as this, girls have another place for validation, support, and friendship.

READING TIME: 3 hours, about 185 pages
THEMES: abandonment, shyness, withdrawal, loneliness, family relationships, coming of age

Discussion Questions

- How would you describe this book to a friend?
- One of the subjects of this book is shyness, taken to its extreme. Describe Anna's shyness. Discuss whether you believe shyness is a problem to be fixed or a normal stage of growth. Have you ever gone through periods of shyness? Discuss how you dealt with it.
- Anna says: "...whenever someone looks right past me without seeing me, I feel myself infinitely superior to him." Explain what you think Anna means by this. Have you ever preferred not to be seen?
- Anna expresses a longing to be an object rather than a person: "Just think how simple and pleasant it would be to go through life as an

object. An attractive little blue sugar bowl with a painted bird on the lid, for instance, sitting in a patch of sunlight on the breakfast table." What does this longing to be inanimate tell you about Anna? If you could be an object, what would it be?

- What were your thoughts about Anna's mother? About her sisters?
- Anna builds herself a secret hideaway within her own house. What is Anna trying to get away from? Have you ever had a secret hiding place in your home or created a fantasy world in your mind? Why do children often create real or imaginary places?
- In what ways do you identify with some of Anna's feelings?
- How does Anna react to the onset of puberty? In what ways are some of her feelings about puberty normal?
- Anna expresses a common adolescent view when she says: "...your own story always seems unique, your own miseries unlike the miseries suffered by anyone else on the planet. It is hard to recognize your own particular predicament as the common fate of millions." What do you think of this quotation?
- Anna chooses to dress up as a moth when she rejoins her human family and is invited to a costume party. What does this choice tell you about her?
- What does Anna mean when she says: "I will build myself a house out of my own flesh and bones where my frightened child-self can find shelter." How does she do this?
- How do you think Anna will fit into the mainstream of human life?

ABOUT THE AUTHOR: Patrice Kindl lives in upstate New York in an old Victorian house with her husband, Paul. Her titles, *The Woman in the Wall* and *Owl in Love* were both included on the American Library Association's list of Best Books for Young Adults. When she isn't writing, Ms. Kindl works with Capuchin monkeys, training them for an organization called Helping Hands, which pairs the animals with quadriplegics.

Beyond the Book...

HIDEAWAY: Anna builds an actual hideaway for herself where she can escape everyday life. Design, draw, or describe for each other what your own fantasy retreat would look like. Bring your ideas together and create a hideout big enough to hold your discussion in. Talk about why it is sometimes comforting to seclude one's self in this way. What's pleasant about feeling isolated? What's frightening about it?

COSTUME PARTY: Host a costume party like the one that Anna finally attends in the book. Choose a costume that you feel is evocative of you, and keep it a secret. At the party, try to guess who's in which costume. Afterwards talk about why you chose the costume you did. Also, discuss how it feels to be disguised. Do you feel like yourself? Why do you think Anna was able to come out while she was wearing a costume?

REFRESHMENTS OR FOOD MENTIONED IN THE BOOK: Anna survives on leftovers from her family's meals. You might enjoy sharing favorite recipes for leftovers, then organize a potluck dinner of delicious "left-overs" as part of your book discussion.

IF YOU LIKED THIS BOOK, TRY...

The Borrowers, by Mary Norton—In many ways, this whimsical novel of "living behind the scenes" parallels some of the themes and situations in *The Woman in the Wall* (see p. 19).

The Stone-Faced Boy, by Paula Fox—A shy boy in a large, boisterous family hides his true feelings behind a "stone face."

Some Other Books by Patrice Kindl:

Owl in Love

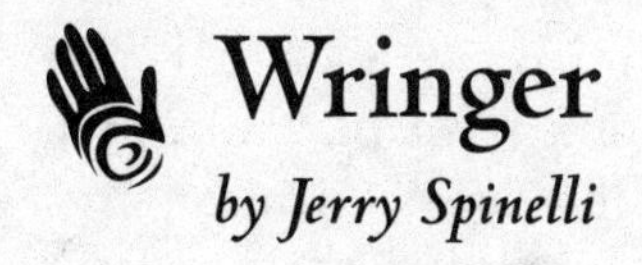

Wringer

by Jerry Spinelli

In this harrowing and suspenseful novel, nine-year-old Palmer knows that two horrible rites of passage await him. Palmer must stoically endure The Treatment, a painful beating at the hands of the town bully. And the second, more wrenching rite, is his expected participation in the killing of wounded birds at his town's annual pigeon shoot. The book explores Palmer's painful dilemma: whether to go along with bullies or put himself at their mercy by defending a friend and a beloved bird.

Peer pressure and bullying are important things for girls this age to be talking about. At one of our book club meetings one mother admitted that she had been a bully while growing up, and she shared her story with her daughter and the group. The girls then spoke about conflicts they had had, some of which their mothers were hearing about for the first time.

READING TIME: 3–4 hours, about 240 pages. Readers should note that the subject of cruelty to animals is treated very graphically in this book.

THEMES: peer pressure, bullying, identity, loyalty, gender roles, cruelty to animals, parent-child relationships

Discussion Questions

- Palmer LaRue seems both thrilled and sickened by the behavior of his new pals Beans, Mutto, and Henry. Why do you think their acceptance is so important to him?
- Why does Palmer's mother intensely dislike Palmer's new friends? How do you feel when your parents comment on your friends?
- Palmer goes along with so many things his friends suggest: teasing Dorothy, hating animals, and behaving obnoxiously. What do you think Palmer is really like?

- ✦ Both Palmer's mother and Dorothy stand on one side of the pigeon-shoot issue while Palmer's father and most of the town's men and boys stand on the other side. What are the female attitudes about this event? How are they different from the accepted male attitudes?
- ✦ Why does Palmer hide so much from his parents? Do you agree with what he does?
- ✦ If Palmer doesn't like to do everything his friends are doing, why is it so hard for him to say no? In your experience, why do you think it is so difficult for a sensible, caring person to stand up against bullies? What's the best way for someone to handle a neighborhood or school bully?
- ✦ Unlike the boys' other bullied victims, Dorothy ignores nearly everything they do. How do they respond to her reaction? Do you believe this is the best way to deal with bullies?
- ✦ Why does Dorothy hold Palmer more accountable for her pain than the other boys who torment her?
- ✦ There are a number of difficult and painful scenes in this book. Which were the hardest for you to read? How did you feel while you were reading them?
- ✦ Palmer proudly survives The Treatment and believes he's had the most perfect birthday a boy can have. Yet that night, when he sees his face in the mirror, he starts crying and can't stop. What is he really crying about?
- ✦ Why does Palmer decide to confide in Dorothy about secretly harboring Nipper? Do you think Palmer could have developed a similar friendship with another boy?
- ✦ As Palmer and Nipper bond more and more, life becomes increasingly painful for Palmer. Why?
- ✦ A powerful line in the book describes Palmer's painful realization that "...it is never the pigeon, but the boy, who is lost." In what ways is Palmer a lost boy?
- ✦ In what ways does Palmer find his truest self by the end of the book?

ABOUT THE AUTHOR: (see p. 187)

Beyond the Book...

PIGEONS: In this book, Jerry Spinelli puts to rest many myths about pigeons, that they are a nuisance, dirty, and unintelligent. Research and share information about these interesting birds. Take a trip to a local park where pigeons or other birds frequent, and feed them stale bread crumbs or birdseed.

BIRTHDAYS: Palmer's ninth and tenth birthdays are the significant events of his life. Share memories of (or plans for) your ninth or tenth birthday. Try to particularly recall or discuss what privileges came or will come with these birthdays. Celebrate your memories and plans with a shared birthday cake and candles.

ROLE-PLAYING: Morgan has a wellness class at school where they do role playing exercises around issues of bullying and peer pressure. Have the girls split up into groups of two or three and invent scenarios that involve peer pressure. Then have them act out the scenes and discuss them during a group meeting.

REFRESHMENTS OR FOOD MENTIONED IN THE BOOK: Serve barbecued chicken, with cake and ice cream for dessert.

IF YOU LIKED THIS BOOK, TRY...

Maniac Magee, by Jerry Spinelli—This Newbery Medal novel portrays the vivid, painful, and energetic life of another memorable young boy (see p. 186).

The Hundred Dresses, by Eleanor Estes—The painful consequences of childhood bullying and teasing are the subject of this classic picture book (see p. 139).

Some Other Books by Jerry Spinelli:

Crash

Jason and Marceline

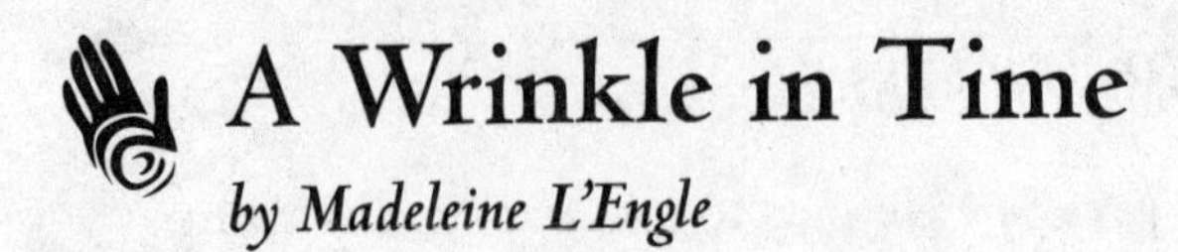

A Wrinkle in Time

by Madeleine L'Engle

This science fiction/fantasy classic is a coming-of-age story and a beat-the-clock page-turner. Meg discovers her previously untapped abilities and hidden strengths when she is forced to rescue her beloved father and younger brother from a sinister force.

While this is a powerful adventure story, the characters are fully developed and believable, and the main point of the book is not the science but the relationships. I'd love to think that my daughter is as brave as Meg, and loves me enough to risk her life to rescue me. Meg is a wonderful role model for girls. She is brave, confident, adventurous, and caring.

READING TIME: 3–4 hours, about 184 pages. Some of the scientific concepts that are discussed or introduced may need to be explained at greater length for clarity.
THEMES: conformity, individuality, young love, family, self-esteem, intelligence

Discussion Questions

- ✦ Does Meg Murry feel comfortable at school? Why or why not? How does she feel about being in the lowest section of her grade?
- ✦ Why are Meg's relationships with her teachers and fellow students so difficult? Have you ever been in similar circumstances? How did that make you feel?
- ✦ Everyone in her small village seems to know that Meg and her family haven't seen or heard from her father in almost a year. How do the gossip and rumors about his absence affect Meg's life in the community?
- ✦ How does Meg feel about having a mother "who was a scientist and a beauty as well"? What do you think it would be like to have as accomplished a mother as Mrs. Murry?

- How does Meg feel about her father and his work?
- Sandy and Dennys, Meg's twin brothers, have no problems fitting into the community. What makes it easier for them than for Meg? How does she feel about the twins?
- How is Charles Wallace like Meg? How is he different?
- Imagine living in a community that mistrusts and resents you. What is it like for the Murrys to live in a community that doesn't understand them?
- Why does Calvin feel like he's in the wrong family? Do you ever feel this way?
- Do Meg and Calvin seem to like each other? How can you tell? What are their similarities and differences?
- Meg's father once tells her not to worry about Charles Wallace because "there's nothing the matter with his mind. He just does things in his own way and in his own time." What is he trying to tell Meg? What is the significance of this situation later in the book?
- When Meg receives her mission from Mrs. Who, Mrs. Which, and Mrs. Whatsit to rescue her father from the dark forces of evil on the planet Camazotz, Mrs. Whatsit bestows upon her a gift that Meg doesn't exactly understand. Mrs. Whatsit says, "Meg, I give you your faults," and explains that they'll come in handy on Camazotz. What does she mean? Do they? Where else would these "faults" come in handy?
- How does Meg conquer her fear of tesseracting? Would you want to do it?
- Can you see any ways in which the collective consciousness on Camazotz is similar to the way people in Meg's village think? Is it similar to the way people in your community think?
- After Charles Wallace is captured by IT, he tells Meg, "On Camazotz, individuals have been done away with. Camazotz is one mind. It's IT. And that's why everybody's so happy and efficient." How does Meg respond? How do you feel about this comment?
- What is significant about the way Meg is finally able to rescue Charles Wallace?

- Why is Meg angry at her father when they are reunited? What does she learn through her experience that allows her to forgive him?
- When she is fighting off the power of IT, Meg has a revelation that, "Like and equal are entirely different things." What does she mean by this? How is this important in the situation? How is this important in life in general?
- How does this adventure change Meg's sense of herself and her abilities?
- What do you think will happen with the relationship between Meg and Calvin after this story ends?

ABOUT THE AUTHOR: Now in her 80s, Newbery Medal winner Madeleine L'Engle has written over 40 books. Children and young-adult readers enjoy her fantasy, science fiction, and family stories, and are especially delighted when familiar characters reappear in different books and series. In addition to *A Wrinkle in Time's* Time Fantasy series, Ms. L'Engle has written the Austin Family series, which includes *Meet the Austins* and *A Ring of Endless Light.* In 1998, Ms. L'Engle received the Margaret A. Edwards Award in honor of her lifetime contribution to young-adult literature.

Beyond the Book...

PLANETARIUM: In the book, Meg, Calvin, and Charles Wallace journey through space. To help imagine what it would be like to take such a journey, visit the local planetarium, or just do some research on your own into the stars and planets in our solar system.

PUMPKIN PLANETS: The planet that captures Meg's father is very different from our world. Using a pumpkin and some acrylic paints or markers, create your own planet. Come up with a name for it, or for its life forms or nations.

TESSERACT: At your local library or on the Internet, do some research on tesseracts and the fourth dimension. Share your findings with each other.

EXPLORATION: Meg's father was experimenting with different dimensions when he was captured by IT. You can experiment with dimensions, too. Talk about what the difference between two-dimensional and three-dimensional objects is. Explore the idea of fourth and fifth dimensions, what they could be and where moving along them could take you.

MOVIE: For older readers, rent a version of *Invasion of the Body Snatchers.* Compare the ideas about conformity and collective consciousness in the film to those in the book.

REFRESHMENTS OR FOOD MENTIONED IN THE BOOK: At the beginning of the book, Meg fixes a sandwich of tuna fish with sweet pickles for Mrs. Whatsit. Make yourself some tuna fish with sweet pickles, or hot cocoa.

IF YOU LIKED THIS BOOK, TRY...

A Wind in the Door; A Swiftly Tilting Planet; Many Waters, by Madeleine L'Engle—These follow the further adventures of the Murry and O'Keefe families.

The Yearling

by Marjorie Kinnan Rawlings

For Jody, a young boy growing up in the Florida backwoods, raising an orphaned fawn offers the companionship and solace absent from the harsh realities of his daily life. His newfound idyllic youth can't last, however, and the boy has to grow up when the fawn does, with bittersweet results.

READING TIME: 5–6 hours, about 428 pages. The regional dialect spoken by characters and the author's descriptive language of fauna and flora may make this somewhat difficult to read.
THEMES: responsibility, loneliness, coming of age, friendship, sacrifice, loyalty, death, family

Discussion Questions

✦ What does it mean for Jody to get something of his own? Why is it so important to him?

✦ Life is hard in the Florida backwoods, and Jody and his family have to scrape and scramble for everything they have. What are some of the rewards of this way of life?

✦ How does Jody relate to nature?

✦ Even though Jody doesn't have what is considered a proper education, what important things has he learned from his experiences?

✦ What values has Jody absorbed from his father, Penny Baxter? How is his relationship with his father different from his relationship with his mother?

✦ Jody observes a neighbor, Buck Forrester, who "was worse than his mother to take away pleasure." How does someone "take away pleasure"? Have you experienced that in your life?

✦ The crippled neighbor boy, Fodder-wing, is Jody's only friend at the beginning of the book. What does Jody learn from his friendship with Fodder-wing? How does Fodder-wing's death change Jody?

✦ Jody's mother seems to care a lot about "duty." What does it mean to her? To Jody? What role does duty play in your life?

✦ Why does Jody identify with the fawn whose mother was shot? Why do you think his parents react so differently to having the fawn come to live with the family? How do these responses reflect their different natures?

✦ Imagine having to hunt and kill your meals, like Jody and his family do. How would you feel about your food?

✦ How would you describe Grandma Hutto? Why is she important to Jody? Is there anyone outside your family who matters to you the way she matters to Jody?

✦ Why is Jody upset by the fight between the Forresters and Oliver Hutto? How does he deal with it? What would you do if you got caught in the middle of a fight between friends?

✦ Jody realizes that "Flag had eased a loneliness that had harassed him, in the very heart of his family." How can someone be lonely when there are people all around? How does Flag change this?

✦ When Flag destroys crops, Jody's father asks Jody to shoot the yearling. How does this change Jody's feelings about his father?

✦ Penny tells Jody, "Life knocks a man down and he gits up and it knocks him down agin." What is he trying to tell his son?

✦ What made the decision about whether or not to shoot Flag so hard for Jody? Have you ever had to make a choice as difficult as this one? What was it, and how did you feel?

✦ How is Jody different at the end of the book than at the beginning? What has he lost, and what has he gained?

ABOUT THE AUTHOR: Marjorie Kinnan Rawlings did most of her writing on manual typewriters in her house in Cross Creek, FL, a house that is now part of a state historic site. She was born in Washington in 1896, graduated Phi Beta Kappa from the University of Wisconsin, and worked as a newspaper reporter. She went to Florida with her then-husband Charles Rawlings in 1928, and started writing there in 1931. *The Yearling* and her other works depict life in rural Florida in the 1930s. A self-professed introvert, she nonetheless enjoyed friendships with fellow authors F. Scott Fitzgerald and Ernest Hemingway. Ms. Kinnan Rawlings died at the age of 57 in 1953.

Beyond the Book...

PRESENTATIONS: Life in the rural backwoods, whether of Florida or elsewhere in the United States at the time *The Yearling* takes place, was often harsh and difficult, and required ingenuity and resourcefulness. Before the group meets, have each mother-daughter pair research and prepare a short presentation on different aspects of daily life during this era, and what kinds of efforts were needed to sustain a family. Talk about the aspects of modern life we take for granted, like refrigeration, communications, transportation, and how our lives would be different if we had to live in Jody's world.

NATURE WALK: Jody's knowledge of animals, and how the natural world works, is invaluable. Using a basic nature guide as a reference, take a walk through a nearby forest or nature preserve. See how many kinds of animal tracks you can identify.

REFRESHMENTS OR FOOD MENTIONED IN THE BOOK: Ordinary snack foods, like popcorn and peppermint candy, were rare delicacies in Jody's life. Serve these treats as refreshment during the discussion.

MOVIE: Rent the video *Old Yeller,* a story about a boy and his dog, where choices that are similar to the decisions Jody had to make are central to the plot.

IF YOU LIKED THIS BOOK, TRY...

Rascal, by Sterling North—This story focuses on a raccoon that becomes a friend and companion.

Girl of the Limberlost, by Eleanor Porter—This contains equally evocative descriptions of the pleasures and perils of backwoods life.

Some Other Books by Marjorie Kinnan Rawlings:

Cross Creek
South Moon Under
Short Stories

Yolanda's Genius

by Carol Fenner

A transplanted city girl learns to translate her Chicago street savvy into friendlier terms when she and her family move to a quiet, suburban community. She champions her misunderstood younger brother so that he may gain wider recognition for his talents in this compelling story.

READING TIME: 2–3 hours, about 211 pages
THEMES: family, love, relationships, race, revenge, friendship, pride, tolerance

Discussion Questions

- Yolonda gets teased a lot by the kids at her new school because of her size. How does that make her feel? Have you ever been teased because of a physical characteristic?
- What kind of relationship does Yolanda have with her brother, Andrew? What's special about it?
- What's interesting about Andrew's relationship with music? How does he use it in his everyday life? How is his music important to Yolanda?
- Andrew is classified as a "special education" student because he has problems with reading. How does Yolanda feel about Andrew's abilities?
- What does it mean to be a "genius" to Yolanda? What does it mean to her classmates? Her teacher? What does it mean to you?
- Yolanda is captivated, during her research, by John Hersey's definition of genius: "True genius rearranges old material in a way never seen before." What do you think this means? Why is this definition so moving to Yolanda? According this definition, who do you consider to be a genius?

- ✦ The harmonica that Andrew plays is the only gift that remained to him from his late father. How does its destruction by the older boys affect Andrew?
- ✦ Yolanda and her mother see Chicago quite differently. How does Yolanda feel about having to leave Chicago? Have you ever had to move from a familiar place? What was it like for you?
- ✦ Why is Shirley's friendship important to Yolanda?
- ✦ Yolanda recognizes that Andrew's music "was part of him, like his hands and his mouth, like his ears. . . .It was his power." What do you think she means? What's your "power"?

ABOUT THE AUTHOR: Carol Fenner was an author before she could write. She remembers composing poems before she was five years old. She would recite them to her mother who would write them down. By the time she was 11, she wrote plays that she and her sister and brothers would perform for neighbors and friends in her basement. Ms. Fenner was also inspired by her aunt, Phyllis Fenner, a writer, who supplied her with books and stories. Carol Fenner still loves to read aloud and often reads her new stories to her family before they're published.

Beyond the Book. . .

JAZZ: Music and jazz are important to Andrew. Gather a wide variety of jazz music (try the local library if you need to). Listen to the music, and during your discussion, talk about how it makes you feel and what images it suggests. As an activity, provide paper and crayons to sketch pictures while you're listening to the music.

CAKE: Yolanda invites Shirley over to her house to bake a cake. Talk about other icebreakers used to initiate friendships, and what activities you think are most successful in getting a friendship off the ground. Bake your favorite cake to share during your discussion.

HAIRSTYLING: Yolanda and her mother have fun when Yolanda's Aunt Tiny comes to visit from Chicago and treats them to a special hairstyling. Have a hairstyling party, with mothers doing their daughters' hair, or vice versa.

IF YOU LIKED THIS BOOK, TRY…

A Wrinkle in Time, by Madeleine L'Engle—In this book, a devoted older sister rescues her misunderstood younger brother from a different sort of prison (see p. 314).

Some Other Books by Carol Fenner:

Randall's Wall
The Skates of Uncle Richard

INDEX